GRAPHING CALCULATOR MANUAL

DARRYL NESTER
Bluffton University

COLLEGE ALGEBRA
TENTH EDITION

Margaret L. Lial
American River College

John Hornsby
University of New Orleans

David I. Schneider
University of Maryland

PEARSON

Addison
Wesley

Boston San Francisco New York
London Toronto Sydney Tokyo Singapore Madrid
Mexico City Munich Paris Cape Town Hong Kong Montreal

PREFACE

This graphing calculator manual was written to help owners of Texas Instruments graphing calculators (TI-83/83+/84+, TI-86, and TI-89) use them to solve problems from *College Algebra*, by Lial, Hornsby, and Schneider. Owners of other TI calculators, like the TI-80, -81, -82, -85, and -92, as well as owners of other brands of graphers, may find the information useful as well, but certainly some of the specific details will not translate directly. In particular, TI-82 users should find many comments in the TI-83/83+/84+ chapter that apply to their calculator, TI-85 users can find useful information in the TI-86 chapter, and TI-92 users should read the chapter on the TI-89.

Please contact me with any questions or corrections concerning this material. My web site also contains additional resources for using graphing calculators; I welcome suggestions as to what else would be useful.

This manual was created with TI calculators (of course), the TI-Connect software, Adobe Photoshop, and TEX (Textures from Blue Sky Research).

Darryl Nester
Bluffton University
Bluffton, Ohio
nesterd@bluffton.edu
http://www.bluffton.edu/~nesterd
October 2007

TABLE OF CONTENTS

This manual contains three chapters, each devoted to one family of calculators. The table below lists, for each calculator, the topics covered in the introductory section (which contains a brief introduction to using the calculator), followed by the examples from *College Algebra*, which are discussed in this manual. The page numbers in parentheses in the first column—e.g., "Section R.2 Example 2 (page 10)"—refer to the text, while the page numbers in the other three columns (under "83/83+/84+," etc.) refer to this manual. For example, information about the TI-89 begins on page 63.

— *continued* —

— continued —

– continued –

Introduction

The information in this section is essentially a summary of material that can be found in the TI-83 manual. Consult that manual for more details. **All references in this chapter to the TI-83 also apply to the TI-83+ and the TI-84+, including the "silver" editions of those calculators.**

While the TI-82 and TI-83 differ in some details, in most cases the instructions given in this chapter can be applied (perhaps with slight alteration) to a TI-82. The icon 82 is used to identify significant differences between the two, but some differences (e.g., a slight difference in keystrokes between the two calculators) are not noted. TI-82 users should watch for these comments. Also, see page 30 for information on computing with complex numbers on the TI-82.

1   Power

To power up the calculator, simply press the ON key. This should bring up the "home screen"—a flashing block cursor, and possibly the results of any previous computations that might have been done.

If the home screen does not appear, one may need to adjust the contrast (see the next section).

To turn the calculator off, press 2nd ON (note that the "second function" of ON—written in yellow type above the key—is "OFF"). The calculator will automatically shut off if no keys are pressed for several minutes.

2   Adjusting screen contrast

If the screen is too dark (all black), decrease the contrast by pressing 2nd then pressing and holding ▼. If the screen is too light, increase the contrast by pressing 2nd and then press and hold ▲.

As one adjusts the contrast, the numbers 1 through 9 will appear in the upper right corner of the screen. If the contrast setting reaches 8 or 9, or if the screen never becomes dark enough to see, the batteries should be replaced.

3   Replacing batteries

To replace the four AAA batteries, first turn the calculator off (2nd ON), then remove the back cover, remove and replace each battery, replace the back cover, then turn the calculator on again. (After replacing batteries, one may need to adjust the contrast down as described above.)

4   Basic operations

Simple computations are entered in essentially the same way they would be written. For example, to compute $2 + 17 \times 5$, press $\boxed{2}\boxed{+}\boxed{1}\boxed{7}\boxed{\times}\boxed{5}\boxed{\text{ENTER}}$ (the $\boxed{\text{ENTER}}$ key tells the calculator to act on what has been typed). Standard order of operations (including parentheses) is followed.

```
2+17*5
          87
■
```

The result of the most recently entered expression is stored in Ans, which is typed by pressing $\boxed{\text{2nd}}\boxed{(-)}$ (the word "ANS" appears in yellow above this key). For example, $\boxed{5}\boxed{+}\boxed{\text{2nd}}\boxed{(-)}\boxed{\text{ENTER}}$ will add 5 to the result of the previous computation.

```
2+17*5
          87
5+Ans
          92
■
```

After pressing $\boxed{\text{ENTER}}$, the TI-83 automatically produces Ans if the first key pressed is one which requires a number before it; the most common of these are $\boxed{+}$, $\boxed{-}$, $\boxed{\times}$, $\boxed{\div}$, $\boxed{\wedge}$, $\boxed{x^{-1}}$, $\boxed{x^2}$, and $\boxed{\text{STO}\blacktriangleright}$. For example, $\boxed{+}\boxed{5}\boxed{\text{ENTER}}$ would accomplish the same thing as the keystrokes above (that is, it adds 5 to the previous result).

```
2+17*5
          87
5+Ans
          92
Ans+5
          97
■
```

Pressing $\boxed{\text{ENTER}}$ by itself evaluates the previously typed expression again. This can be especially useful in conjunction with Ans. The screen on the right shows the result of pressing $\boxed{\text{ENTER}}$ a second time.

```
2+17*5
          87
5+Ans
          92
Ans+5
          97
         102
■
```

Several expressions can be evaluated together by separating them with colons ($\boxed{\text{ALPHA}}\boxed{.}$). When $\boxed{\text{ENTER}}$ is pressed, the result of the *last* computation is displayed. The screen shown illustrates the computation $2(5+1)^2$.

```
3+2
           5
Ans+1:Ans²:2Ans
          72
■
```

5   Cursors

When typing, the appearance of the cursor indicates the behavior of the next keypress. When the standard cursor (a flashing solid block, ■) is visible, the next keypress will produce its standard action—that is, the command or character printed on the key itself.

If $\boxed{\text{2nd}}\boxed{\text{DEL}}$ is pressed, the TI-83 is placed in INSERT mode and the standard cursor will appear as a flashing underscore. If the arrow keys ($\boxed{\blacktriangle}$, $\boxed{\blacktriangledown}$, $\boxed{\blacktriangleright}$, $\boxed{\blacktriangleleft}$) are used to move the cursor around within the expression, and the TI-83 is placed in INSERT mode, subsequent characters and commands will be inserted in the line at the cursor's position. When the cursor appears as a block, the TI-83 is in DELETE (or OVERWRITE) mode, and subsequent keypresses will replace the character or command at the cursor's position. (When the cursor is at the end of the expression, this is irrelevant.)

The TI-83 will return to DELETE mode when any arrow key is pressed. It can also be returned to DELETE mode by pressing $\boxed{\text{2nd}}\boxed{\text{DEL}}$ a second time.

Pressing $\boxed{\text{2nd}}$ causes an arrow to appear in the cursor: ◨ (or an underscored arrow). The next keypress will produce its "second function"—the command or character printed in yellow above the key. (The cursor will then return to "standard.") If $\boxed{\text{2nd}}$ is pressed by mistake, pressing it a second time will return the cursor to standard.

Pressing ALPHA places the letter "A" in the cursor: ▣ (or an underscored "A"). The next keypress will produce the letter or other character printed in green above that key (if any), and the cursor will then return to standard. Pressing ALPHA a second time cancels ALPHA mode. Pressing 2nd ALPHA "locks" the TI-83 in ALPHA mode, so that all of the following keypresses will produce characters until ALPHA is pressed again, or until some menu or second function is accessed.

6   Accessing previous entries ("deep recall")

By repeatedly pressing 2nd ENTER ("ENTRY"), previously typed expressions can be retrieved for editing and re-evaluation. Pressing 2nd ENTER once recalls the most recent entry; pressing 2nd ENTER again brings up the second most recent, etc. The number of previous entries thus displayed varies with the length of each expression (the TI-83 allocates 128 bytes to store previous expressions).

7   Menus

Keys such as WINDOW, MATH and VARS bring up a menu screen with a variety of options. The top line of the menu screen gives a collection of submenus (if any), which can be selected with the ◄ and ► keys. The lower lines list the available commands; these can be selected using the ▲ and ▼ keys and ENTER, or by pressing the number (or letter) preceding the desired option. Shown is the menu produced by pressing MATH; the arrow next to the 7 in the bottom row indicates that there are more options available below.

The various commands in these menus are too numerous to be listed here. They will be mentioned as needed in the examples.

8   Variables

The letters A through Z can be used as variables (or "memory") to store numerical values. To store a value, type the number (or an expression) followed by STO►, then a letter (preceded by ALPHA if necessary), then ENTER. That letter can then be used in the same way as a number, as demonstrated at right.

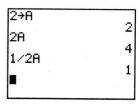

Note: The TI-83 interprets 2A as "2 times A"—the "*" symbol is not required (this is consistent with how we interpret mathematical notation). As for order of operations, this kind of multiplication is treated the same as "*" multiplication.

[82] *This latter comment is **not** true of the TI-82; on the TI-82, implied multiplication (such as 2A) is done before other multiplication and division, and even before some other operations, like the square root function ∫. Therefore, for example, the expression 1/2A is evaluated as 1/4 on the TI-82 (assuming that A is 2).*

9   Setting the modes

By pressing the MODE key, one can change many aspects of how the calculator behaves. For most of the examples in this manual, the "default" settings should be used; that is, the MODE screen should be as shown on the right. Each of the options is described below; consult the TI-83 manual for more details. Changes in the settings are made using the arrows keys and ENTER. (Note: The TI-84+ has essentially the same screen, but also includes a line for setting the clock.)

The Normal Sci Eng setting specifies how numbers should be displayed. The screen on the right shows the number 12345 displayed in Normal mode (which displays numbers in the range $\pm 9,999,999,999$ with no exponents), Sci mode (which displays all numbers in scientific notation), and Eng mode (which uses only exponents that are multiples of 3). Note: "E" is short for "times 10 to the power," so $1.2345\text{E}4 = 1.2345 \times 10^4 = 1.2345 \times 10000 = 12345$.

The Float 0123456789 setting specifies how many places after the decimal should be displayed. The default, Float, means that the TI-83 should display all non-zero digits (up to a maximum of 10).

Radian Degree indicates whether angle measurements should be assumed to be in radians or degrees. (A right angle measures $\frac{\pi}{2}$ radians, which is equivalent to $90°$.) This text does not refer to angle measurement.

Func Par Pol Seq specifies whether formulas entered into the Y= screen are functions (specifically, y as a function of x), parametric equations (x and y as functions of t), polar equations (r as a function of θ), or sequences (u, v and w as functions of n). The text accompanying this manual uses the first and last of these modes.

When plotting a graph, the Connected Dot setting tells the TI-83 whether or not to connect the individually plotted points. Sequential Simul specifies whether individual expressions should be graphed one at a time (sequentially), or all at once (simultaneously).

Real a+bi re^θi specifies how to deal with complex numbers. Real means that only real results will be allowed (unless i is entered as part of a computation)—so that, for example, taking the square root of a negative number produces an error ("NONREAL ANS"). Selecting one of the other two options means that square roots of negative numbers are allowed, and will be displayed in "rectangular" ($a + bi$) or polar ($re^{i\theta}$) format. The text uses only the first of these formats. More information about complex numbers can be found beginning on page 12 (Section 1.3, Example 1) of this manual.

[82] *The TI-82 does not support complex numbers, so it does not include this setting. However, some complex computations can be done with a TI-82; see the appendix at the end of this chapter, page 30.*

Finally, the Full Horiz G-T setting allows the option of, for example, showing both the graph and the home screen, as in the screen on the right (this shows a "horizontal split").

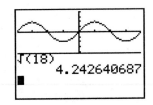

The third option, G-T, has no effect on the home screen display, but will show graphs and tables side by side when GRAPH is pressed.
[82] *The TI-82 supports only the horizontal split.*

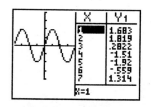

A related group of settings are found in the FORMAT menu (2nd ZOOM). The default settings are shown in the screen on the right, and are generally the best choices for most examples in this book (although the last two settings could go either way). [82] *The TI-82 does not include the* ExprOn ExprOff *option.*

RectGC PolarGC specifies whether graph coordinates should be displayed in rectangular (x, y) or polar (r, θ) format. Note that this choice is independent of the Func Par Pol Seq mode setting. The CoordOn CoordOff setting determines whether or not graph coordinates should be displayed. GridOff GridOn specifies whether or not to display a grid of dots on the graph screen, while AxesOn AxesOff and LabelOff LabelOn do the same thing for the axes and labels (y and x) on the axes. ExprOn ExprOff specifies whether or not to display the formula (expression) of the curves on the GRAPH screen when tracing. (This can be useful when more than one graph is displayed.)

10   Setting the graph window

The exact contents of the WINDOW menu vary depending on whether the calculator is in function, parametric, polar, or sequence mode; below are four examples showing the WINDOW menu in each of these modes.

WINDOW	WINDOW	WINDOW	WINDOW
Xmin=-4.7	Tmin=0	θmin=0	nMin=4
Xmax=4.7	Tmax=6.2831853…	θmax=6.2831853…	nMax=10
Xscl=1	Tstep=.1308996…	θstep=.1308996…	PlotStart=1
Ymin=-3.1	Xmin=-4.7	Xmin=-4.7	PlotStep=1
Ymax=3.1	Xmax=4.7	Xmax=4.7	Xmin=-4.7
Yscl=1	Xscl=1	Xscl=1	Xmax=4.7
Xres=1	↓Ymin=-3.1	↓Ymin=-3.1	↓Xscl=1
Function mode	Parametric mode	Polar mode	Sequence mode

All these menus include the values Xmin, Xmax, Xscl, Ymin, Ymax, and Yscl. When the GRAPH key is pressed, the TI-83 will show a portion of the Cartesian (x-y) plane determined by these values. In function mode, this menu also includes Xres, the behavior of which is described in section 12 of this manual (page 7). The other settings in the WINDOW screen allow specification of the smallest, largest, and step values of t (for parametric mode) or θ (for polar mode), or initial conditions for sequence mode.

With settings as in the example screens shown above, the TI-83 would display the screen at right: x values from -4.7 to 4.7 (that is, from Xmin to Xmax), and y values between -3.1 to 3.1 (Ymin to Ymax). Since Xscl = Yscl = 1, the TI-83 places tick marks on both axes every 1 unit; thus the x-axis ticks are at -4, -3, $\ldots$, 3, and 4, and the y-axis ticks fall on the integers from -3 to 3. This window is called the "decimal" window, and is most quickly set by pressing ZOOM 4.

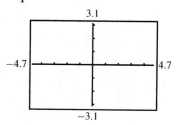

Below are four more sets of WINDOW settings, and the graph screens they produce. Note that the first graph on the left has tick marks every 10 units on both axes. The second window is called the "standard" viewing window, and is most quickly set by pressing $\boxed{\text{ZOOM}}\boxed{6}$. The setting $\text{Yscl} = 0$ in the final graph means that no tick marks are placed on the y-axis.

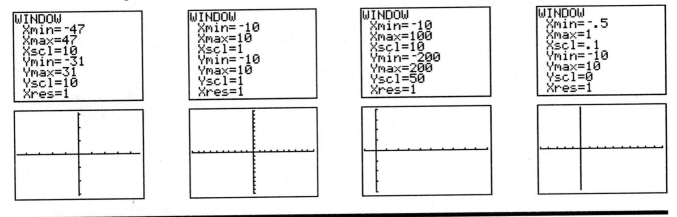

11   The graph screen

The TI-83 screen is made up of an array of square dots (pixels) with 63 rows and 95 columns. All the pixels in the leftmost column have x-coordinate Xmin, while those in the rightmost column have x-coordinate Xmax. The x-coordinate changes steadily across the screen from left to right, which means that the coordinate for the nth column (counting the leftmost column as column 0) must be $\text{Xmin} + n\Delta\text{X}$, where $\Delta\text{X} = (\text{Xmax} - \text{Xmin})/94$. Similarly, the nth row of the screen (counting up from the bottom row, which is row 0) has y-coordinate $\text{Ymin} + n\Delta\text{Y}$, where $\Delta\text{Y} = (\text{Ymax} - \text{Ymin})/62$.

Note: In (horizontal) split screen mode, $\Delta\text{Y} = (\text{Ymax} - \text{Ymin})/30$. In G–T (vertical split screen) mode, $\Delta\text{X} = (\text{Xmax} - \text{Xmin})/46$ and $\Delta\text{Y} = (\text{Ymax} - \text{Ymin})/50$.

It is not necessary to memorize the formulas for ΔX and ΔY. Should they be needed, they can be determined by pressing $\boxed{\text{GRAPH}}$ and then the arrow keys. When pressing $\boxed{\blacktriangleright}$ or $\boxed{\blacktriangleleft}$ successively, the displayed x-coordinate changes by ΔX; meanwhile, when pressing $\boxed{\blacktriangle}$ or $\boxed{\blacktriangledown}$, the y-coordinate changes by ΔY. Alternatively, the values can be found by pressing $\boxed{\text{VARS}}\boxed{1}\boxed{8}\boxed{\text{ENTER}}$ (for ΔX) or $\boxed{\text{VARS}}\boxed{1}\boxed{9}\boxed{\text{ENTER}}$ (for ΔY). This produces results like those shown on the right.

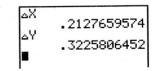

In the decimal window $\text{Xmin} = -4.7$, $\text{Xmax} = 4.7$, $\text{Ymin} = -3.1$, $\text{Ymax} = 3.1$, note that $\Delta\text{X} = 0.1$ and $\Delta\text{Y} = 0.1$. Thus, the individual pixels on the screen represent x-coordinates $-4.7, -4.6, -4.5, \ldots, 4.5$, $4.6, 4.7$ and y-coordinates $-3.1, -3, -2.9, \ldots, 2.9, 3, 3.1$. This is where the decimal window gets its name.

Windows for which $\Delta\text{X} = \Delta\text{Y}$, such as the decimal window, are called square windows. Any window can be made square be pressing $\boxed{\text{ZOOM}}\boxed{5}$. To see the effect of a square window, observe the two pairs of graphs below. In each pair, the first graph is on the standard window, and the second is on a square window (after pressing $\boxed{\text{ZOOM}}\boxed{5}$). The first pair shows the lines $y = 2x - 3$ and $y = 3 - \frac{1}{2}x$; note that on the square window, these lines look perpendicular (as they should). The second pair shows a circle centered at the origin with a radius of 8. On the standard window, this looks like an oval since the screen is wider than it is tall. (The reason for the gaps in the circle will be addressed in the next section.)

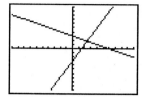 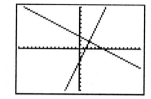 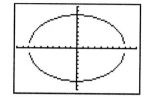 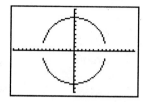

12   Graphing a function

This introductory section only addresses creating graphs in function mode. This textbook does not include examples of parametric and polar graphs; however, procedures for creating such graphs are very similar.

To see the graph of $y = 2x - 3$, begin by entering the formula into the calculator. This is done on the $\boxed{Y=}$ screen of the calculator. Select one of the variables Y_1, Y_2, ..., and enter the formula. If other y variables have formulas, either erase them (by positioning the cursor on that line and pressing $\boxed{CLEAR}$) or position the cursor on the equals sign "=" for that line and press $\boxed{ENTER}$ (this has the effect of "unhighlighting" the equals sign, which tells the TI-83 not to graph that formula). Additionally, if any of Plot1, Plot2 or Plot3 are highlighted, move the cursor up until it is on that plot and press $\boxed{ENTER}$. In the screen on the right, only Y_1 will be graphed.

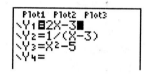

The next step is to choose a viewing window; see the previous section for more details on this. This example uses the standard window ($\boxed{ZOOM}\boxed{6}$).

Finally, press $\boxed{GRAPH}$, and the line should be drawn. In order to produce this graph, the TI-83 considers 95 values of x, ranging from Xmin to Xmax in steps of $\triangle$X (assuming that Xres $= 1$; see below for other possibilities). For each value of x, it computes the corresponding value of y, then plots that point (x, y) and (if the calculator is in Connected mode) draws a line between this point and the previous one.

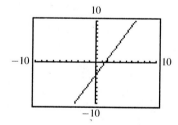

If Xres is set to 2, the TI-83 will only compute y for every other x value; that is, it uses a step size of $2\triangle$X. Similarly, if Xres is 3, the step size will be $3\triangle$X, and so on. Setting Xres higher causes graphs to appear faster (since fewer points are plotted), but for some functions, the graph may look "choppy" if Xres is too large, since detail is sacrificed for speed.

Note: If the line does not appear, or the TI-83 reports an error, double-check all the previous steps. Also, check the mode settings (discussed in section 9, page 4).

Once the graph is visible, the window can be changed using $\boxed{WINDOW}$ or $\boxed{ZOOM}$. Pressing the $\boxed{TRACE}$ key brings up the "trace cursor," and displays the x- and y-coordinates for various points on the line as the $\boxed{\triangleleft}$ and $\boxed{\triangleright}$ keys are pressed. Tracing beyond the left or right columns causes the TI-83 to adjust the values of Xmin and Xmax and redraw the graph.

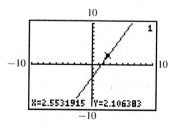

To graph the function

$$y = \frac{1}{x-3},$$

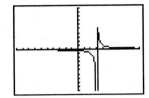

enter that formula into the Y= screen (note the use of parentheses). As before, this example uses the standard viewing window.

When GRAPH is pressed, the TI-83 produces the graph shown on the right. This illustrates one of the pitfalls of the connect-the-dots method used by the calculator: The nearly-vertical line segment drawn at $x = 3$ *should not be there*, but it is drawn because the calculator connects the points

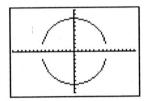

$$x = 2.9787234, \ y = -46.99999 \ \text{ and } \ x = 3.1914894, \ y = 5.2222223.$$

Calculator users must learn to recognize these flaws in calculator-produced graphs. **Note:** Recent versions of the TI-84+ operating system "fix" this problem, so that this vertical line is not displayed.

The graph of a circle centered at the origin with radius 8 (shown on the square window ZOOM 6 ZOOM 5) shows another problem that arises from connecting the dots. When $x = -8.064516$, y is undefined, so no point is plotted (that is, there is no point on this circle that has x-coordinate less than -8, or greater than 8). The next point plotted on the upper half of the circle is $x = -7.741935$ and $y = 2.0155483$; since no point had been plotted for the previous x-coordinate, this is not connected to anything, so there appears to be a gap between the circle and the x-axis. The calculator is not "smart" enough to know that the graph should extend from -8 to 8.

One additional feature of graphing with the TI-83 is that each function can have a "style" assigned to its graph. The symbol to the left of Y_1, Y_2, etc. indicates this style, which can be changed by pressing ◀ until the cursor is over the symbol, then pressing ENTER to cycle through the options. These options are shown on the right (with brief descriptive names); complete details are in the TI-83 manual. [82] *The TI-82 does not include graph-style features.*

13 Adding programs to the TI-83

The TI-83's capabilities can be extended by downloading or entering programs into the calculator's memory. Instructions for writing a program are beyond the scope of this manual, but programs written by others and downloaded from the Internet (or obtained as printouts) can be transferred to the calculator in one of three ways:

1. If one TI-83 already has a program, it can be transferred to another using the calculator-to-calculator link cable. To do this, first make sure the cable is firmly inserted in both calculators. On the sending calculator, press 2nd X,T,Θ,n (LINK), then 3, and then select (by using the ▲ and ▼ keys and ENTER) the program(s) to be transferred. Now press the ▶ to bring up the TRANSMIT submenu. *Before* pressing ENTER on the sending calculator, prepare the receiving calculator by pressing 2nd X,T,Θ,n ▶ ENTER, and *then* press ENTER on the sending calculator.

2. If a computer with the TI-Graph Link is available, and the program file is on that computer (e.g., after having been downloaded from the Internet), the program can be transferred to the calculator using the TI Connect (or TI Graph Link) software. This transfer is done in a manner similar to the calculator-to-calculator transfer described above; specific instructions can be found in the documentation that

accompanies the software. (They are not given here because of slight differences between platforms and software versions.)

3. View a listing of the program and type it in manually. (**Note:** Even if the TI-Graph Link cable is not available, the software can be used to view program listings on a computer.) While this is the most tedious method, studying programs written by others can be a good way to learn programming. To enter a program, start by choosing PRGM ◄ 1 ("Create New"), then type a name for the new program (like "QUADFORM" or "MIDPOINT")—note that the TI-83 is automatically put into ALPHA mode. Then type each command in the program, and press 2nd MODE (QUIT) to return to the home screen when finished.

To run the program, make sure there is nothing on the current line of the home screen, then press PRGM, select the number or letter of the program (a sample screen is shown), and press ENTER. If the program was entered manually (option 3 above), errors may be reported; in that case, choose GOTO, correct the mistake and try again.

Programs can be found at many places on the Internet, including:

- http://www.bluffton.edu/~nesterd—the Web site of the author of this manual;

- http://tifaq.calc.org—A "Frequently Asked Questions" page maintained by Ray Kremer; and

- http://www.ticalc.org.

Additionally, the TI-83+ and TI-84+ calculators include a variety of "APPs"—applications (programs) which can extend the capabilities of the calculator in various ways. APPs can be viewed by selecting the APPS key; the number of installed APPs depends on the model. Shown is the list of applications available on the TI-84+ Silver Edition. Additional APPs can be downloaded from education.ti.com, then installed using a Graph Link cable.

Examples

Here are the details for using the TI-83 for several of the examples from the textbook. Also given are the keystrokes necessary to produce some of the commands shown in the text's examples. In some cases, some suggestions are made for using the calculator more efficiently.

Throughout this section, it is assumed that the textbook is available for reference. The problems from the text are not restated here, and there are frequent references to the calculator screens shown in the text.

Section R.2   Example 2   (page 10)   Evaluating Exponential Expressions

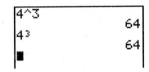

This example discusses how to evaluate 4^3, $(-6)^2$, -6^2, $4 \cdot 3^2$, and $(4 \cdot 3)^2$. To evaluate 4^3, one could type [4][^][3][ENTER], or a superscripted 3 can be produced by pressing [4][MATH][3][ENTER]. Of course, the results are the same either way. (Note: Only reciprocating, squaring, and cubing—that is, the powers $-1, 2$, and 3—can be displayed as superscripts. All other powers must use the [^] key.)

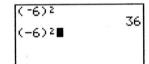

All other computations in this example are straightforward with the TI-83, except for this *caveat*: For $(-6)^2$ and -6^2, one must use the [(-)] key rather than the [-] key to enter "negative 6." In the screen shown, the first line was typed as [(][(-)][6][)][x^2], while the second line was [(][-][6][)][x^2]. When [ENTER] is pressed, the second line produces a syntax error—the calculator's way of saying that the line makes no sense.

Section R.2   Example 3   (page 11)   Using Order of Operations

Using the calculator for parts (a) and (b) of this example is straightforward; the expressions are entered as printed. For (c) and (d), we have the expressions

$$\frac{4 + 3^2}{6 - 5 \cdot 3} \quad \text{and} \quad \frac{-(-3)^3 + (-5)}{2(-8) - 5(3)}.$$

Entering these on the TI-83 requires us to add some additional parentheses to make sure the correct order of operations is followed.

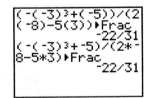

Proper entry of the expression in (d) is shown on the right, in two different forms. The second illustrates some time-saving that can be done in keying in such expressions: The parentheses around -5 in the numerator were omitted, and the denominator $2(-8) - 5(3)$ was typed in as [2][×][(-)][8][-][5][×][3]—the parentheses were exchanged for the multiplication symbol. It would be fine to type the expression unchanged, but the extra parentheses make it somewhat harder to read (and also mean more opportunities to make a mistake).

The screen shown above also illustrates the command ▸Frac. This command (which simply means "display the result of this computation as a fraction, if possible") is available as [MATH][1].

Section R.2   Example 7   (page 15)   Evaluating Absolute Value

Section R.2   Example 9   (page 16)   Evaluating Absolute Value Expressions

The absolute value function is typed using MATH ▸ 1, and displays as abs(. Expressions like those in these examples would be entered as abs(-5/8) or abs(-8+2) on the TI-83. Note that the TI-83 automatically supplies the opening parenthesis.

82 *On the TI-82, the absolute value function is typed using* 2nd x⁻¹, *and the opening parenthesis is* not *included. Parentheses are optional for expressions like* -abs -2, *but are needed for* abs (-8+2).

Section R.6   Example 4   (page 55)   Using the Definition of $a^{1/n}$

In evaluating these fractional exponents with a calculator, some time can be saved by entering fractions as decimals. The screen on the right performs the computations for (a), (b) and (c) with fewer keystrokes than entering, for example, 36^(1/2). (Of course, as noted in Section R.7 of the text, $36^{1/2} = \sqrt{36}$, which requires even fewer keystrokes.)

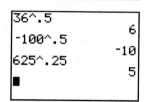

Note for (d), the parentheses around -1296 cannot be omitted: The text observes that this is not a real number, but the calculator will display a real result if the parentheses are left out. (See also the next example.) For (e), the decimal equivalent of $1/3$ is $0.\overline{3}$, which can be entered as a sufficiently long string of 3s (at least 12). The last line shows a better way to do this; the TI-83's order of operations is such that 3^{-1} is evaluated before the other exponentiation. This method would also work for the other computations; e.g., (a) could be entered 3 6 ^ 2 x⁻¹.

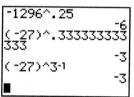

Section R.6   Example 5   (page 56)   Using the Definition of $a^{m/n}$

For (f), attempting to evaluate $(-4)^{5/2}$ will produce either a NONREAL ANS error, or a complex number result, depending on the Real a+bi re^θi mode setting, described on page 4.

82 *On the TI-82, the only possible result is a* NONREAL ANS *error, as the TI-82 does not support complex results.*

Section R.7   Example 1   (page 63)   Evaluating Roots

Aside from the square root function, produced with 2nd x², the cube root (³√ ()) and "x-root" (ˣ√) functions are available as MATH 4 and MATH 5. For example, the input 4ˣ√ (16) computes $\sqrt[4]{16}$. It is worth noting that fourth roots can be typed in more efficiently using two square roots, since $\sqrt[4]{a} = a^{1/4} = (a^{1/2})^{1/2} = \sqrt{a^{1/2}} = \sqrt{\sqrt{a}}$.

Section 1.1   Example 1   (page 85)   Solving a Linear Equation

The TI-83 offers several approaches to solving (or confirming a solution to) an
equation. Aside from graphical approaches to solving equations (which are discussed later, on page 18), here are some non-graphical (numerical) approaches:

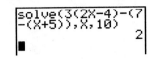

As illustrated on the right, the TI-83's `solve` function attempts to find a value of X that makes the given
expression equal to 0, given a guess (10, in this case).

The `solve` command is accessed through the function catalog: Press `2nd` `0`, then `4` (which brings up the
commands beginning with "t"). Press `▲` until the triangle cursor (▸) points to `solve(`, then press `ENTER`.
For the equations in this section, the "guess" value does not affect the result, but for equations in later
sections, different guesses might produce different results. Full details on how to use this function (found
in the CATALOG) can be found in the TI-83 manual.

Since it is difficult to access this command, it would be a good idea to either use deep recall (see page 3) if
`solve` is needed several times in a row, or to use the "interactive solver" built-in to the TI-83. This feature
is found by pressing `MATH` `0`.

If Solver has not been used before, the first screen below (a) appears, asking for an expression involving at
least one variable. The TI-83 will attempt to find values of that variable which make the expression equal
to 0. The expression for this example is shown in screen (b). Pressing `ENTER` brings up screen (c). Type
the guess (this example uses 5), then press `ALPHA` `ENTER` and the TI-83 seeks a solution near 5, which is
reported in screen (d).

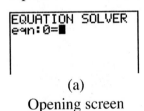

| (a) | (b) | (c) | (d) |
| Opening screen | Enter expression | Enter guess | Solution |

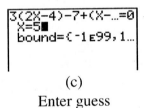

If Solver *has* been used before, `MATH` `0` jumps directly to screen (c). To change the expression, press `▲`.

Note: In this example, we learned how the TI-83 can be used to support an analytic solution. But the
TI-83 and any other graphing calculator also can be used for solving problems when an analytic solution
is **not** possible—that is, when one cannot solve an equation "algebraically." This is often the case in many
"real-life" applications, and is one of the best arguments for the use of graphing calculators.

82 *The TI-82 does not include this interactive solver; the* `solve` *function is accessed with* `MATH` `0`.

Section 1.3   Example 1   (page 104)   Writing $\sqrt{-a}$ as $i\sqrt{a}$
Section 1.3   Example 2   (page 105)   Finding Products and Quotients Involving Negative Radicands
Section 1.3   Example 3   (page 105)   Simplifying a Quotient Involving a Negative Radicand

The TI-83 will not do these computations unless it is first put in `a+bi` mode; see page 4. (One could also
use `re^θi` mode, but this would report complex numbers in "polar format," rather than the format used in
the text.) See also the comment related to Figure 4 on page 104 of the text.

[82] *The TI-82 will not do computations with complex results. See the appendix to the chapter (page 30) for information about how to "fake" these computations.*

Section 1.3 Example 4 (page 106)	Adding and Subtracting Complex Numbers
Section 1.3 Example 5 (page 107)	Multiplying Complex Numbers
Section 1.3 Example 6 (page 107)	Simplifying Powers of *i*
Section 1.3 Example 7 (page 108)	Dividing Complex Numbers

The character " i " is 2nd . . Although not completely necessary, it is a good idea to put the TI-83 in a+bi mode; see page 4. (Even in Real mode, the TI-83 will display results for computations in which i is entered directly; it only complains if asked to find even roots of negative numbers.)

[82] *See the appendix to the chapter (page 30) for information about doing these computations on a TI-82.*

Section 1.5   Example 1   (page 122)   Solving a Problem Involving the Volume of a Box

A *table* can be a useful tool to solve equations like this. To use the table features of the TI-83, begin by entering the formula ($y = 15x^2 - 200x + 500$) on the Y= screen, as one would to create a graph. (The highlighted equals signs determine which formulas will be displayed in the table, just as they do for graphs.)

Next, press 2nd WINDOW to access the TABLE SETUP screen. The table will display y values for given values of x. The TblStart value sets the lowest value of x, while ΔTbl determines the "step size" for successive values of x. These two values are only used if the Indpnt option is set to Auto—this means, "automatically generate the values of the independent variable (x)." The effect of setting this option to Ask is illustrated at the end of this example. (The Depend option should almost always be set to Auto; if it is set to Ask, the y values are not displayed until ENTER is pressed.)

When the TABLE SETUP options are set satisfactorily, press 2nd GRAPH to produce the table. Shown (above, right) is the table generated based on the settings in the above screen; the fact that y is negative for $x = 8$ and $x = 9$ supports the restriction $x > 10$ from Step 3 of the example. By pressing ▼ repeatedly, the x values are increased, and the y values updated. (Pressing ▲ decreases the x values, but clearly that is not appropriate for this problem.) After pressing ▼ eleven times, the table looks like the second screen, which shows that Y1 equals 1435 when X equals 17.

Extension: Suppose the specifications called for the box to be 1500 cubic inches (instead of 1435). We see from the table above that x must be between 17 and 18 inches. The TI-83's table capabilities provide a convenient way to "zoom in" on the value of x (which could also be found with the solve command, or by other methods).

Since $17 < x < 18$, set TblStart to 17 and ΔTbl to 0.1. This produces the table shown on the right, from which we see that $17.2 < x < 17.3$. Now set TblStart to 17.2 and ΔTbl to 0.01, which reveals that x is between 17.2 and 17.21. This process can be continued to achieve any desired degree of accuracy.

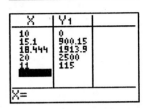

Finally, the screen on the right shows the effect of setting Indpnt to Ask. Initially, the table is blank, but as values of x are entered, the y values are computed. Up to seven x values can be entered; if more are desired, the [DEL] key can be used to make room.

| Section 1.7 | Example 1 | (page 146) | Solving a Linear Inequality |
| Section 1.7 | Example 2 | (page 147) | Solving a Linear Inequality |

Shown is one way to visualize the solution to inequalities like these. The inequality symbols $>$, $<$, $\geq$, $\leq$ are found in the TEST menu ([2nd][MATH]). The TI-83 responds with 1 when a statement is true and 0 for false statements. This is why the graph of Y1 appears as it does: For values of x less than 4, the inequality is true (and so Y1 equals 1), and for $x \geq 4$, the inequality is false (and Y1 equals 0). Note that this picture does *not* help one determine what happens when $x = 4$.

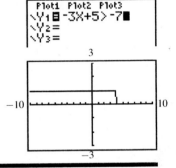

| Section 1.7 | Example 3 | (page 148) | Solving a Three-Part Inequality |

To see a "picture" of this inequality like the one shown for the previous two examples requires a little additional work. Setting Y1=-2<5+3X<20 will *not* work correctly. Instead, one must enter either

    Y1=(-2<5+3X)(5+3X<20)   or   Y1=(-2<5+3X) and (5+3X<20)

which will produce the desired results: A function that equals 1 between $-\frac{7}{3}$ and 5, is equals 0 elsewhere. ("and" is found in the TEST:LOGIC menu, [2nd][MATH][▶].)

Section 1.7	Example 5	(page 149)	Solving a Quadratic Inequality
Section 1.7	Example 6	(page 150)	Solving a Quadratic Inequality
Section 1.7	Example 8	(page 152)	Solving a Rational Inequality
Section 1.7	Example 9	(page 153)	Solving a Rational Inequality

The methods described above can also be applied to these inequalities.

Extension: Can the calculator be used to solve the inequality $x^2 + 6x + 9 > 0$? Yes—if it is used carefully. Shown is the graph of the expression $x^2 + 6x + 9 > 0$. This suggests that the inequality is true for all x; that is, the solution set is $(-\infty, \infty)$.

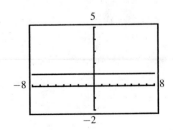

This conclusion is **not correct**: The graph of $y = x^2 + 6x + 9$ (without ">0") touches the x-axis at $x = -3$, which means that this must be excluded from the solution set, since ">0" means the graph must be above (not on) the x-axis. The correct solution set is $(-\infty, -3) \cup (-3, \infty)$.

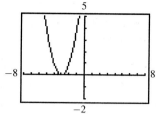

This example should serve as a warning: Sometimes, the TI-83 will mislead you. When possible, try to confirm the calculator's answers in some way.

Section 1.8   Example 1   (page 159)   Solving Absolute Value Equations

Section 1.8   Example 2   (page 160)   Solving Absolute Value Inequalities

The "solver" features described on page 12 of this manual can be used for these equations, too. Recall that abs(is entered as MATH ▶ 1 .

Section 2.1   Example 5   (page 186)   Using the Midpoint Formula

The TI-83 can do midpoint computations nicely by putting coordinates in a list— that is, using braces (2nd (and 2nd)) instead of parentheses. When adding two lists, the calculator simply adds corresponding elements, so the two x-coordinates are added, as are the y-coordinates. Dividing by 2 completes the task.

Section 2.2   Example 2   (page 194)   Graphing Circles

The text suggests graphing Y1=4+√(36-(X+3)²) and Y2=4-√(36-(X+3)²). Here are three options to speed up entering these formulas:

- After typing the formula in Y1, move the cursor to Y2 and press 2nd STO▶ (RCL), then VARS ▶ 1 1 ENTER . This will "recall" the formula of Y1, placing that formula in Y2. Now edit this formula, changing the first "+" to a "−."

- After typing the formula in Y1, enter Y2=8-Y1. (To type Y1, press VARS ▶ 1 1 .) This produces the desired results, since $8 - Y_1 = 8 - (4 + \sqrt{36 - (x+3)^2}) = 8 - 4 - \sqrt{36 - (x+3)^2} = 4 - \sqrt{36 - (x+3)^2}$.

- Enter the single formula Y1=4+{-1,1}√(36-(X+3)²). (The curly braces { and } are 2nd (and 2nd)). When a list (like {-1,1}) appears in a formula, it tells the TI-83 to graph this formula several times, using each value in the list.

The window chosen in the text is a square window (see section 11 of the introduction of this chapter), so that the graph looks like a circle. (On a non-square window, the graph would look like an ellipse—that is, a squashed circle.) Other square windows would also produce a "true" circle, but some will leave gaps similar to those circles shown in sections 11 and 12 of the introduction (pages 6–7). Shown is the same circle on the square window produced by pressing ZOOM 6 ZOOM 5 .

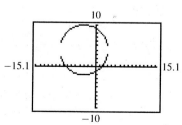

Section 2.3 Example 6 (page 208)	Using Function Notation
Section 2.3 Example 7 (page 209)	Using Function Notation

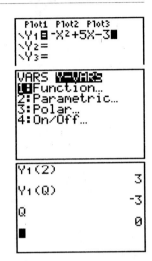

The TI-83 will do *some* computations in function notation. Simply enter the function in Y₁ (or any other function). Return to the home screen, and press [VARS] [▶] (Y-VARS). Option 1 allows access to all the function (y) variables; parametric and polar variables are also available (as options 2 and 3). Select the appropriate function (Y₁ in this case), then type (2) to evaluate that function with the input 2. The result shown on the screen agrees with (a) in the text.

The TI-83 is less helpful if asked to evaluate expressions like those given in Example 6(b) or Example 7. Note, for example, that Y₁(Q) produces the result -3, rather than the desired result $-Q^2 + 5Q - 3$. When asked to evaluate an expression using a variable (like Q), the TI-83 simply substitutes the current value of that variable, which in this case was 0.

Section 2.4   Example 3   (page 219)   Graphing a Vertical Line

The text notes that such a line is not a function; therefore, it cannot be graphed by entering the equation on the [Y=] screen. However, the TI-83 does provide a fairly straightforward method of graphing vertical lines using the Vertical command.

From the home screen, press [2nd][PRGM][4] to access the Vertical command, then type the x-coordinate of the line. The command shown on the right was typed as [2nd][PRGM][4][(-)][3]; when [ENTER] is pressed, the line will be drawn.

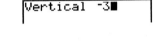

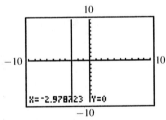

Alternatively, the vertical line can be drawn "interactively" if the Vertical command is issued from the graph screen. This allows the user to move the cursor (and the vertical line) to the desired position using the [◀] and [▶] keys. Pressing [ENTER] places the line at that location, then allows more lines to be drawn. (Press [GRAPH] to finish drawing vertical lines.) Note that since $x = -3$ does not correspond to a column of pixels on the screen (see page 6), the vertical line is actually located at $x \approx -2.9787$, but the appearance on the screen is the same as if it were at $x = -3$.

Section 2.5   Example 6   (page 236)   Finding Equations of Parallel and Perpendicular Lines

In the discussion following this example (on page 237), the text notes that the lines from part (b) do not appear to be perpendicular unless they are plotted on a square window. See section 11 of the introduction of this chapter (page 6) for more information about square windows.

Aside from the methods described in these two examples, the text refers to the technique of linear regression on page 241. The steps are illustrated in Figure 51; here is a more detailed description of the process (using the data of Example 8).

Given a set of data pairs (x, y), the TI-83 can find various formulas (including linear and quadratic, as well as more complex formulas) that approximate the relationship between x and y. These formulas are called "regression formulas."

The first step in determining the regression formula is to enter the data into the TI-83. This is done by pressing STAT , then choosing option 1 (Edit).

This brings up the list-editing screen. (If the column headings are not L_1, L_2, L_3 as shown on the right, press STAT 5 ENTER to reset the statistics [list] editor to its default. Then return to the list-editing screen.) Enter the year values into the first column (L_1) and the percentage of women in the labor force into the second column (L_2). If either column already contains data, the DEL key can be used to delete numbers one at a time, or—to delete the whole column at once—press the ▲ key until the cursor is at the top of the column (on L_1 or L_2) and press CLEAR ENTER . Make sure that both columns contain the same number of entries.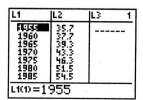

When all data is entered, press STAT ▶ 4 to choose LinReg(ax+b) from the CALC statistics submenu. This will place the command LinReg(ax+b) on the home screen.

To tell the TI-83 where to find x and y, press 2nd 1 , 2nd 2 , which adds "L_1,L_2" to the command. (**Note:** This is not absolutely necessary, since the TI-83 will assume that x is in L_1 and y is in L_2 unless told otherwise. In other words, "LinReg(ax+b)" (by itself) works the same as "LinReg(ax+b) L_1,L_2".) Pressing ENTER should then produce the second screen shown in Figure 51 of the text. (It may also include a value of "R^2," but this can be ignored.)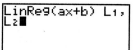

If it produces an error message, it will likely either be a DIM MISMATCH (meaning that the two lists L_1 and L_2 have different numbers of entries) or a SYNTAX error, probably because the LinReg(ax+b) command was not on a line by itself. The screen on the right, for example, produces a syntax error.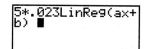

Section 2.5   Example 9   (page 241)   Solving an Equation with a Graphing Calculator

The text illustrates a graphical process for solving the equation $-2x - 4(2-x) = 3x + 4$; in this discussion, we will show two graphical methods for solving the equation $\frac{1}{2}x - 6 = \frac{3}{4}x - 9$. (The answer is $x = 12$.)

The first method is the **intersection** method. To begin, set up the TI-83 to graph the left side of the equation as Y1, and the right side as Y2. **Note:** Putting the fractions in parentheses ensures no mistakes with order of operations. This is not crucial for the TI-83, but is a good practice because some other models give priority to implied multiplication. See section 8 of the introduction, page 3.

We are looking for an x value that will make the left and right sides of this equation equal to each other, which corresponds to the x-coordinate of the point of intersection of these two graphs.

Next, select a viewing window which shows the point of intersection; we use $[-15, 15] \times [-10, 10]$ for this example. The TI-83 can automatically locate this point using the CALC menu ([2nd][TRACE]). Choose option 5 (intersect), use [▲], [▼] and [ENTER] to specify which two functions to use (in this case, the only two being displayed), and then use [◀] or [▶] to specify a guess. After pressing [ENTER], the TI-83 will try to find an intersection of the two graphs. The screens below illustrate these steps.

[2nd][TRACE][5]
selects the
CALC:intersect feature

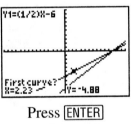

Press [ENTER]
to choose Y1

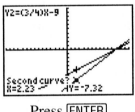

Press [ENTER]
to choose Y2

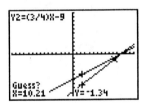

Move cursor to
specify guess and
then press [ENTER].

The final result of this process is the screen shown on the right. The x-coordinate of this point of intersection is calculated to 14 digits of accuracy, so if the solution were some less "convenient" number (say, $\sqrt{3}$ or $1/\pi$), we would have an answer that would be accurate enough for nearly any computation.

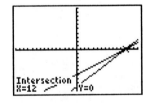

Note: An approximation for the point of intersection can be found simply by moving the TRACE cursor as near the intersection as possible. The amount of error can be minimized by "zooming in" on the graph. This was the only method available for early graphing calculators like the TI-81.

The second graphical approach is to use the x-**intercept method**, which seeks the x-coordinate of the point where a graph crosses the x-axis. Specifically, we want to know where the graph of Y1−Y2 crosses the x-axis, where Y1 and Y2 are as defined above. This is because the equation $\frac{1}{2}x - 6 = \frac{3}{4}x - 9$ can only be true when $\frac{1}{2}x - 6 - \left(\frac{3}{4}x - 9\right) = 0$. (This is the approach illustrated in the text for Example 9.)

To find this x-intercept, begin by defining Y3=Y1−Y2 on the Y= screen. We could do this by re-typing the formulas entered for Y1 and Y2, but having typed those formulas once, it is more efficient to do this as shown on the right. To type Y1 and Y2, press VARS ▶ 1 to access the Y-VARS:FUNCTION menu. Note that Y1 and 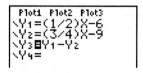 Y2 have been "de-selected" so that they will not be graphed (see section 12 of the introduction, page 7).

We must first select a viewing window which shows the x-intercept; we again use $[-15, 15] \times [-10, 10]$. The TI-83 can automatically locate this point by choosing option 2 from the CALC (2nd TRACE) menu. The TI-83 prompts for left and right bounds and a guess, then attempts to locate the zero between the given bounds. (Provided there is only one zero between the bounds, and the function is "well-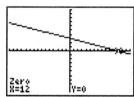 behaved"—meaning it has some nice properties like continuity—the calculator will find it.) The screens below illustrate these steps.

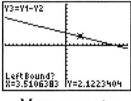   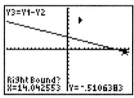   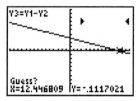

| Move cursor to the left of the zero, press ENTER | Move cursor to the right of the zero, press ENTER | Move cursor close to the zero, press ENTER | The TI-83 finds the zero. |

Section 2.6   Example 2   (page 252)   Graphing Piecewise-Defined Functions

The text shows two methods for entering piecewise-defined functions on the TI-83. Recall that the inequality symbols $>$, $<$, $\geq$, $\leq$ are found in the TEST menu (2nd MATH). The use of DOT mode is not crucial to the graphing process, as long as one remembers that vertical line segments connecting the "pieces" of the graph (in this case, at $x = 2$) are not really part of the graph.

The advantage of the method used in (b)—placing the piecewise definition in a single formula—is that Y1 will act exactly like the function f. For example, to evaluate $f(5)$ in (a), one would first have to check which formula to use (Y1 or Y2). To evaluate $f(5)$ in (b), simply enter Y1(5). Also, when a single formula is used, TRACE will properly show the behavior of the function as the trace cursor moves left and right.

Extension: For more complicated piecewise-defined functions, a similar procedure can be used. Consider

$$f(x) = \begin{cases} 4 - x^2 & \text{if } x < -1 \\ 2 + x & \text{if } -1 \leq x \leq 4 \\ -2 & \text{if } x > 4 \end{cases}$$

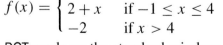

shown on the right in DOT mode on the standard window.

It might be tempting to enter Y1=(4−X²)(X<−1)+(2+X)(−1≤X≤4)+(−2)(X>4), but this does not work. Instead, use either

    Y1=(4−X²)(X<−1)+(2+X)(−1≤X and X≤4)+(−2)(X>4),   or

    Y1=(4−X²)(X<−1)+(2+X)(−1≤X)(X≤4)+(−2)(X>4)

(the "and" in the first formula is found using 2nd MATH ▶ 1).

Section 2.6   Example 3   (page 253)   Graphing a Greatest Integer Function

We wish to graph the step function $y = [\![\frac{1}{2}x + 1]\!]$. This is entered on the TI-83 as shown on the right (either Y_1 or Y_2). int(is found in the MATH:NUM menu, MATH ▶ 5.

Section 2.7   Example 1   (page 259)   Stretching or Shrinking a Graph
Section 2.7   Example 2   (page 261)   Reflecting a Graph Across an Axis
Section 2.7   Example 6   (page 265)   Translating a Graph Vertically
Section 2.7   Example 7   (page 266)   Translating a Graph Horizontally

See page 8 for information on setting the thickness of a graph. Note that the symbol next to Y_2 in the calculator screens shown in the text indicates that the graph should be thick.

It is possible to distinguish between two graphs without having them drawn using different styles. When two or more graphs are drawn, the TRACE feature can be used to determine which graphs correspond to which formulas: By pressing ▲ and ▼, the trace cursor jumps from one graph to another, and the upper right corner displays the number of the current formula. On the right, the trace cursor is on graph 2—that is, the graph of $g(x) = 2|x|$.

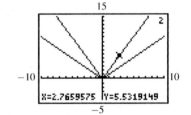

Alternatively, when ExprOn is selected on the TI-83's graph format screen (2nd ZOOM—see page 5), the calculator will display in the upper left corner the formula for the graph currently being traced. The screen on the right illustrates this feature.

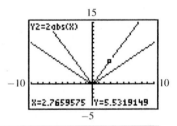

Section 2.8   Example 5   (page 279)   Evaluating Composite Functions

To type Y_1, Y_2, ... on the home screen, press VARS ▶ 1.

Section 3.1   Example 1   (page 304)   Graphing Quadratic Functions
Section 3.1   Example 2   (page 305)   Graphing a Parabola by Completing the Square
Section 3.1   Example 3   (page 306)   Graphing a Parabola by Completing the Square

The TI-83 can automatically locate the vertex of the parabola (to a reasonable degree of accuracy) using the CALC menu (2nd TRACE).

We illustrate this procedure for the function in Example 1(c). After defining $Y_1 = -(1/2)(X-4)^2 + 3$ and choosing a suitable window, press 2nd TRACE and select option 4 for a downward opening parabola, or option 3 for an upward opening parabola. Next, specify left and right bounds (numbers that are, respectively, less than and greater than the location of the vertex) and a guess, and the TI-83 will try to find the highest (or lowest) point on the graph between the given bounds. (For some functions, it might not always find the correct maximum or minimum, but for parabolas, it should succeed.) The screens below illustrate these steps.

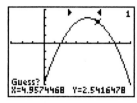
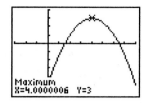

[2nd][TRACE][4]

Move cursor to the left of the zero, press [ENTER]

...move the cursor twice more

...and *voilà!*

Section 3.1   Example 6   (page 310)   Modeling the Number of Hospital Outpatient Visits

See page 17 for information about using the TI-83 for regression computations.

Section 3.2   Example 3   (page 325)   Deciding Whether a Number is a Zero

The TI-83's "function notation" evaluation method (page 16 of this manual) can also be used to check for zeros: In the screen on the right, Y_1, Y_2, and Y_3 have been defined (respectively) as the functions in (a), (b), and (c). While this method fails for complex numbers—$Y_3(1+2i)$ results in an error—one can evaluate Y_3 at $1 + 2i$ by first storing that value in X, then requesting the value of Y_3 (with no parentheses, which means "evaluate using the current value of X"). The screen on the right shows the output of this sequence of commands. While the TI-83 does not report the value as 0, it lends support to the belief that $1 + 2i$ is a zero since both the real and complex parts are very small. (The difference is due to roundoff error.)

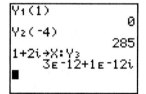

Section 3.3   Example 6   (page 335)   Finding All Zeros of a Polynomial Function Given One Zero

Calculator programs can be downloaded from the Internet to automate the process of synthetic division, and even the determination of zeros for polynomials. See section 13 of the introduction (page 8) for details about calculator programs.

If installed on your calculator (TI-83+ or TI-84+ only), the PolySmlt APP can find all zeros (real and complex) of a polynomial. The screens below illustrate the process. Note in the third screen, the top line contains a reminder that the expression must be equal to 0.

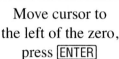
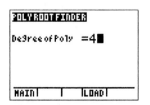
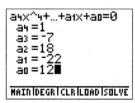

Option 1 finds polynomial roots (zeros).

Give the degree of the polynomial.

Enter coefficients on this screen.

Press [GRAPH] to find the zeros.

Section 3.5   Example 2   (page 361)   Graphing a Rational Function

Note that this function is entered as Y₁=2/(X+1), **not** Y₁=2/X+1.

The issue of incorrectly drawn asymptotes is also addressed in section 12 of the introduction (page 7). Changing the window to Xmin $= -5$ and Xmax $= 3$ (or any choice of Xmin and Xmax which has -1 halfway between them) eliminates this vertical line because it forces the TI-83 to attempt to evaluate the function at $x = -1$. Since f is not defined at -1, it cannot plot a point there, and as a result, it does not attempt to connect the dots across the "break" in the graph.

Section 4.1   Example 7   (page 409)   Finding the Inverse of a Function with a Restricted Domain

Note that Figure 10 shows a square viewing window (see page 6); the mirror-image property of the inverse function would not be as clear on a non-square window.

Section 4.2   Example 11   (page 426)   Using Data to Model Exponential Growth

See page 17 for information about using the TI-83 for regression computations.

Section 4.4   Example 1   (page 448)   Finding pH

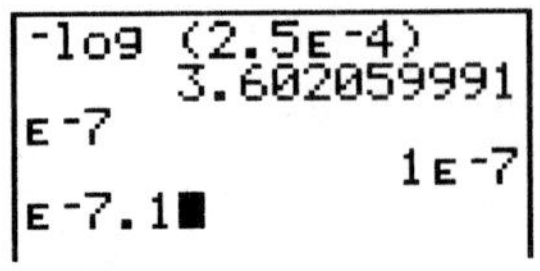

For (a), the text shows -log(2.5*10^(-4)), but this could also be entered as shown on the first line of the screen on the right, since "E" (produced with 2nd ,) and "*10^" are nearly equivalent. The two are not completely interchangeable, however; in particular, in part (b), "10^" **cannot** be replaced with "E", because "E" is only valid when followed by an *integer*. That is, E-7 produces the same result as 10^-7, but the last line shown on the screen produces a syntax error.

(Incidentally, "10^" is 2nd LOG, but 1 0 ^ produces the same results.)

Section 4.5   Example 8   (page 463)   Modeling Coal Consumption in the U.S.

The modeling function given in the text was found using a regression procedure (LnReg, or logarithmic regression) similar to that described on page 17.

Section 5.1   Example 8   (page 502)   Using Curve Fitting to Find an Equation Through Three Points

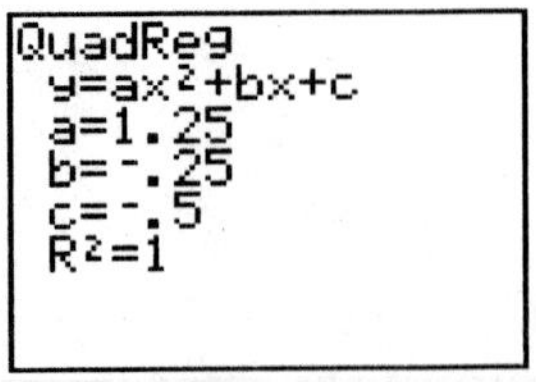

Note that the equation found in this example by algebraic methods can also be found very quickly on the TI-83 by performing a quadratic regression (see page 17) on the three given points.

Section 5.2   Example 1   (page 512)   Using the Gauss-Jordan Method

Section 5.2   Example 2   (page 514)   Using the Gauss-Jordan Method

Note that the TI-82 and TI-83 have a [MATRX] key, while on the TI-83+ and TI-84+, [2nd][x^{-1}] accesses the matrix commands.

Before describing the row operations, a few words about entering matrices into the TI-83: The TI-83 has ten matrices, named [A] through [J]. One way to enter a matrix is by "storing" the contents on the home screen; an example is shown on the right. "[B]" is typed by pressing [MATRX][2] (or [2nd][x^{-1}][2]), while the other square brackets [and] are [2nd][×] and [2nd][−].

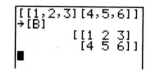

Alternatively, press [MATRX][◄] (or [2nd][x^{-1}][◄]), then choose one of the matrices, specify the number of rows and columns, and type in the entries, moving around by pressing the arrow keys and [ENTER]. The screen on the right shows the process of editing matrix [A]. Press [2nd][MODE] (QUIT) when finished.

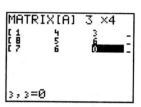

The formats for the matrix row-operation commands, found in the MATRX:MATH menu, are:

- rowSwap(*matrix*,*A*,*B*) produces a new matrix that has row *A* and row *B* swapped.
- row+(*matrix*,*A*,*B*) produces a new matrix with row *A* added to row *B*.
- *row(*number*,*matrix*,*A*) produces a new matrix with row *A* multiplied by *number*.
- *row+(*number*,*matrix*,*A*,*B*) produces a new matrix with row *A* multiplied by *number* and added to row *B* (row *A* is unchanged).

Shown below are examples of each of these operations on the matrix $[E] = \begin{bmatrix} 1 & 4 & 7 \\ 2 & 5 & 8 \\ 3 & 6 & 9 \end{bmatrix}$.

```
rowSwap([E],1,2       row+([E],1,2           *row(-4,[E],1          *row+(2,[E],1,3
        [[2 5 8]             [[1 4 7 ]             [[-4 -16 -28]          [[1  4  7 ]
         [1 4 7]              [3 9 15]              [2   5   8 ]          [2  5  8 ]
         [3 6 9]]             [3 6 9 ]]             [3   6   9 ]]         [5 14 23]]
■                     ■                     ■                      ■
```

Keep in mind that these row operations leave the matrix [E] untouched. To perform a sequence of row operations, each result must either be stored in a matrix, or use the result variable Ans as the matrix. For example, with [A] equal to the augmented matrix for the system given in this example, the following screens illustrate the initial steps in the Gauss-Jordan method. Note the use of Ans in the last two screens.

```
[A]                    *row+(-3,[A],1,2       [0 6  -16 28]          [0 6  -16 28]
 [[1 -1  5  -6]        )                        [1 3   2  5 ]]         [0 4  -3  11]]
  [3  3 -1 10]          [[1 -1 5  -6]         *row+(-1,Ans,1,3       *row(1/6,Ans,2)▶
  [1  3  2  5 ]]         [0  6 -16 28]        )                      Frac
■                        [1  3  2  5 ]]        [[1 -1 5   -6]        [[1 -1 5   -6…
                       ■                         [0  6 -16 28]         [0 1  -8/3 14…
                                                 [0  4 -3  11]]        [0 4  -3  11 …
                                               ■                      ■
```

As we see in Figure 7(d), the rref command —[MATRX][►][ALPHA] (or [2nd][x^{-1}][►][ALPHA]), then [MATRX] (or [APPS])—will do all the necessary row operations at once, making these individual steps seem tedious.

[82] *The TI-82 will perform the individual row operations, but does not have the* rref *command.*

Section 5.3 Example 1 (page 523)	Evaluating a 2×2 Determinant
Section 5.3 Example 3 (page 526)	Evaluating a 3×3 Determinant

To type "det([A])," press MATRX ▶ 1 (or 2nd x^{-1} ▶ 1) (since det is under the MATRX:MATH submenu) then MATRX 1 (or 2nd x^{-1} 1), then) (although this command will work correctly without the closing parenthesis).

The determinant of a 3×3—or larger—matrix is as easy to find with a calculator as that of a 2×2 matrix. (At least, it is as easy for the user; the calculator is doing all the work!) Note, however, that trying to find the determinant of a non-square matrix (for example, a 3×4 matrix) results in an INVALID DIM error.

Section 5.5   Example 2   (page 544)   Solving a Nonlinear System by Elimination

See the discussion of Example 2 from Section 2.2 (page 15 of this manual) for tips on entering formulas like these.

The text shows the intersections as found by the procedure built in to the calculator (described on page 18 of this manual). However, that is somewhat misleading; for these equations, the TI-83 can only find these intersections if the "guesses" supplied by the user are the exact x coordinates of the intersections (that is, -2 and 2). This is because the two circle equations are only valid for $-2 \leq x \leq 2$, while the hyperbola equations are only valid for $x \leq -2$ and $x \geq 2$. Since only ± 2 fall in both of these domains, any guess other than these two values results in a BAD GUESS error.

Section 5.6   Example 1   (page 555)   Graphing a Linear Inequality

With the TI-83, there are two ways to shade above or below a function. The simpler way is to use the "shade above" graph style (see page 8). The screen on the right shows the "shade above" symbol next to Y₁, which produces the graph shown in the text

The other way to shade is the Shade(command, found as option 7 in the 2nd PRGM (DRAW) menu. The format is

>   Shade(*lower*, *upper*, *min X*, *max X*, *pattern*, *resolution*)

Here *lower* and *upper* are the functions between which the TI-83 will draw the shading (above *lower* and below *upper*). The last four options can be omitted. *min X* and *max X* specify the starting and ending x values for the shading. If omitted, the TI-83 uses Xmin and Xmax.

The last two options specify how the shading should look. *pattern* determines the direction of the shading: 1 (vertical—the default), 2 (horizontal), 3 (negative-slope 45°—that is, upper left to lower right), or 4 (positive-slope 45°—that is, lower left to upper right). *resolution* is a positive integer (1,2,3,...) which specifies how dense the shading should be (1 = shade every column of pixels, 2 = shade every other column, 3 = shade every third column, etc.). If omitted, the TI-83 shades every column; i.e., it uses *resolution* = 1.

To produce a graph like the one accompanying Example 1 in the text, the appropriate command (typed on the home screen) would be something like

    Shade(-(1/4)X+1,10,-2,6,1,2).

The use of 10 for the *upper* function simply tells the TI-83 to shade up as high as necessary; this could be replaced by any number greater than 4 (the value of Ymax for the viewing window shown). Also, if the function Y₁ had previously been defined as -(1/4)X+1, this command could be shortened to Shade(Y₁,10,-2,6,1,2).

One more useful piece of information: Suppose one makes a mistake in typing the Shade(command (e.g., switching *upper* and *lower*, or using the wrong value of *resolution*), resulting in the wrong shading. The screen on the right, for example, arose from typing Shade(-(1/4)X+1,1,-2,6,1,2). In order to achieve the desired results, the mistake must be erased using the Clr-Draw command (2nd PRGM 1 ENTER). Then—perhaps using deep recall (see page 3)—correct the mistake in the Shade(command and try again.

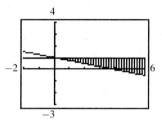

One more approach is available on the TI-83+ and TI-84+ calculators: The Inequalz APP (see section 13 of the introduction, page 8) changes the Y= screen so that one can enter inequalities rather than equations. After entering the formula (Y₁=-(1/4)X+1), place the cursor on the equals sign and press ALPHA ZOOM to choose the ≤ graphing option.

The chief advantage of using this APP is that the calculator attempts to show more detail—specifically, it draws either a solid or a dashed line to indicate that the region satisfying the inequality does or does not include the line. (In this case, the line is solid.)

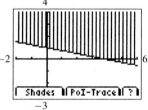

When finished graphing inequalities, select the Inequalz APP once again to tell it to quit. Otherwise, the inequality graphing options will show up every time the Y= screen is accessed.

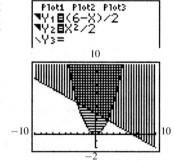

Section 5.6   Example 2   (page 556)   Graphing Systems of Inequalities

The easiest way to produce (essentially) the same graph as that shown in the text is to use the "shade above" graph style (see page 8). The screen on the right (above) shows the "shade above" symbol next to Y₁ and Y₂, with the results shown on the graph below. When more than one function is graphed with shading, the TI-83 rotates through the four shading patterns (see page 24); that is, it graphs the first with vertical shading, the second with horizontal, and so on. All shading is done with a resolution of 2 (every other pixel).

The Shade command can be used to produce this from the home screen. If Y₁=(6-X)/2 and Y₂=X²/2, the commands at right produce the graph shown in the text.

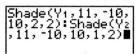

A nicer picture can be created, with a little more work, by making the observation that if y is greater than both $x^2/2$ and $(6-x)/2$, then for any x, y must be greater than the larger of these two expressions. The TI-83 provides a convenient way to find the larger of two numbers with the max(function ($\boxed{\text{2nd}}\boxed{\text{STAT}}\boxed{\triangleleft}\boxed{2}$), so that max(Y₁,Y₂) will return the larger of Y₁ and Y₂, and the functions shown on the $\boxed{\text{Y=}}$ screen on the right (above) will produce the graph shown below it. (Note the graph style settings: Y₁ and Y₂ display as solid curves, while Y₃ has the "shade above" style.) Similar results would result from the home-screen command Shade(max(Y₁,Y₂),11,-10,10,1,2).

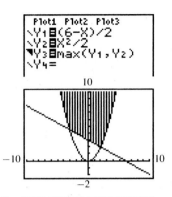

Using the Inequalz APP on the TI-83+ or TI-84+ (see the previous example), a nice picture can be made fairly easily. First, enter Y₁>(6-X)/2 and Y₂> X²/2, which produces a graph much like the first one shown above. Now, pressing $\boxed{\text{ALPHA}}\boxed{\text{Y=}}$ or $\boxed{\text{ALPHA}}\boxed{\text{WINDOW}}$ will bring up the Shades options; if we select Ineq Intersection, we get the desired region.

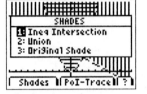 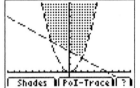

Extension: The table below shows how (using graph styles or home-screen commands) to shade regions that arise from variations on the inequalities in this example, assuming that Y₁=(6-X)/2 and Y₂=X²/2. (The results of these commands are not shown here.)

For the system...	or equivalently...	the command would be...
$x < 6 - 2y$ $x^2 < 2y$	$y < (6-x)/2$ $y > x^2/2$	shade below Y₁ and above Y₂, or enter Shade(Y₂,Y₁,-10,10,1,2)
$x > 6 - 2y$ $x^2 > 2y$	$y > (6-x)/2$ $y < x^2/2$	shade above Y₁ and below Y₂, or enter Shade(Y₁,Y₂,-10,10,1,2)
$x < 6 - 2y$ $x^2 > 2y$	$y < (6-x)/2$ $y < x^2/2$	shade below Y₁ and below Y₂, or enter Shade(-10,min(Y₁,Y₂),-10,10,1,2)

See page 23 for information about matrices on the TI-83.

The identity function is option 5 under the MATRX:MATH submenu. To produce the 3×3 identity matrix, for example, press $\boxed{\text{MATRX}}$ (or $\boxed{\text{2nd}}\boxed{x^{-1}}$), then $\boxed{\triangleright}\boxed{5}\boxed{3}\boxed{)}\boxed{\text{ENTER}}$.

Section 5.8	Example 2	(page 583)	Finding the Inverse of a 3×3 Matrix
Section 5.8	Example 3	(page 584)	Identifying a Matrix with No Inverse

---

An inverse matrix is found using $\boxed{x^{-1}}$. For example, to find the inverse of matrix $[A]$, press $\boxed{\text{MATRX}}\boxed{1}\boxed{x^{-1}}$ $\boxed{\text{ENTER}}$ (or $\boxed{\text{2nd}}\boxed{x^{-1}}\boxed{1}\boxed{x^{-1}}\boxed{\text{ENTER}}$). An error will occur if the matrix is not square, or if it is a singular matrix (as in Example 3).

Section 6.1	Example 1	(page 607)	Graphing a Parabola with Horizontal Axis
Section 6.1	Example 2	(page 608)	Graphing a Parabola with Horizontal Axis
Section 6.2	Example 1	(page 618)	Graphing Ellipses Centered at the Origin
Section 6.3	Example 1	(page 629)	Using Asymptotes to Graph a Hyperbola

---

See the discussion of Example 2 from Section 2.2 (page 15 of this manual) for tips on entering formulas like these.

When entering these formulas, make sure to use enough parentheses, so that operations are performed in the correct order; e.g., for Example 2 of Section 6.1, the formula in Y₁ should look like this:

    Y₁=-1.5+√((X-.5)/2)

Also, when looking at a calculator graph like the one shown in Figure 18 (page 619), keep in mind that the apparent gap between this graph and the x-axis is not really there; it is a flaw that arises from the calculator's method of plotting functions (see discussion on page 8 of this manual).

If installed on your calculator (TI-83+ or TI-84+ only), the Conics APP will graph conic sections, given the appropriate information. Shown below is the process of graphing the parabola $(y + 3)^2 = -8(x - 2)$; other graphs are accomplished in a similar manner. Press $\boxed{\text{GRAPH}}$ to produce the graph, $\boxed{\text{TRACE}}$ to see points on the graph, and $\boxed{\text{Y=}}$ to change the conic parameters. The APP chooses the viewing window for you.

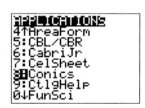

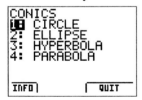 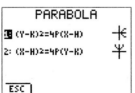 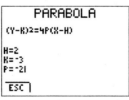 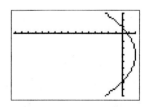

Section 7.1   Example 3   (page 655)   Modeling Insect Population Growth

---

One way to produce this sequence (and similar recursively defined sequences) is with the Ans variable, as shown on the right. After pressing $\boxed{1}\boxed{\text{ENTER}}$, then typing the next line and pressing $\boxed{\text{ENTER}}$, we simply press $\boxed{\text{ENTER}}$ over and over to generate successive terms in the sequence. Each time, Ans is replaced by the previous result.

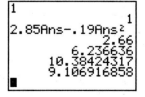

Figure 4 of the text shows a table of values created by putting the TI-83 into Sequence mode, and making the definitions shown on the right. (This screen is accessed with the [Y=] key.) The plot in Figure 5(b) was also created in this way; see the discussion of Example 7 from Section 7.3 (below) for details.

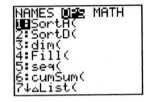

Section 7.1   Example 4   (page 657)   Using Summation Notation

Section 7.2   Example 9   (page 669)   Using Summation Notation

---

The seq(command is item 5 in the [2nd][STAT][▶] (LIST:OPS) menu. Given a formula a_n for the nth term in a sequence, the command

    seq(*formula*, *variable*, *start*, *end*, *step*)

produces the list $\{a_{start}, a_{start+step}, \ldots, a_{end}\}$. In most uses, the value of *step* is 1, which is the assumed value if this is omitted. The resulting list can have no more than 999 items (or possibly fewer, if memory is low).

Note that *variable* can be any letter—K and I are used in the text, but X is more convenient (since it can be typed with [X,T,Θ,*n*]). Also, in each of the commands shown on the right, the final ", 1)" could be omitted.

The sum command found by pressing [2nd][STAT][◀][5]. sum can be applied to any list—either one of the built-in lists L₁,..., L₆, or a list created with the seq(command, or a user-defined list.

Section 7.3   Example 7   (page 676)   Summing the Terms of an Infinite Geometric Series

---

The plot shown in the text can be produced in at least two ways. The first method involves putting the TI-83 in sequence mode, then entering the formula for S_n on the u(n) line of the [Y=] screen, as on the right. Note that the graph style is set to "dotted," based on the symbol to the left of u(n). Also, the value of u(nMin)=u(1) must be specified.

Next, press [WINDOW] and make the settings shown on the right. The viewing window is as shown in the text ($\texttt{Xmin} = 0$, $\texttt{Xmax} = 6$, $\texttt{Ymin} = 0$, $\texttt{Ymax} = 2$).

Finally, press [2nd][ZOOM] and check that the first option (which only appears in sequence mode) is set to Time. Pressing [GRAPH] should produce a plot like that shown in the text.

The second method begins by storing lists in L1 and L2 as shown on the right, so that L1 contains values of n, and L2 contains the corresponding values of S_n. (The slightly shorter command $(1-(1/3)^{\wedge}L_1)/(2/3){\rightarrow}L_2$ has the same effect as the second command shown.) Note that this approach is less appealing for recursively defined sequences.

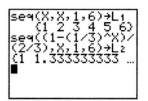

Next press $\boxed{\text{2nd}}\boxed{\text{Y=}}$ and $\boxed{1}$ and make the settings for a STAT PLOT shown on the right.

Finally, set up the viewing window, and check that nothing else will be plotted; that is, press $\boxed{\text{Y=}}$ and make sure that the only thing highlighted is Plot1. Pressing $\boxed{\text{GRAPH}}$ should produce a plot like that shown in the text.

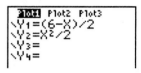

Note: When finished with a STAT PLOT, it is a good idea to turn all statistics plots off so that the TI-83 will not attempt to display them the next time $\boxed{\text{GRAPH}}$ is pushed. This is most easily done by executing the PlotsOff command, using the key sequence $\boxed{\text{2nd}}\boxed{\text{Y=}}\boxed{4}\boxed{\text{ENTER}}$.

Section 7.4   Example 1   (page 687)   Evaluating Binomial Coefficients

Section 7.6   Example 4   (page 701)   Using the Permutations Formula

---

The nCr and nPr commands are found by pressing $\boxed{\text{MATH}}\boxed{\blacktriangleleft}$, then either $\boxed{3}$ or $\boxed{2}$. (Also located in that menu is the factorial operator "!".)

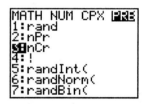

Section 7.7   Example 6   (page 716)   Finding Probabilities in a Binomial Experiment

---

Statistical distribution functions are found by pressing $\boxed{\text{2nd}}\boxed{\text{VARS}}$; binompdf is option 0 in this menu. As the text explains, the format for this function is

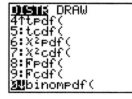

binompdf(*# of trials*, *prob. of success*, *# of successes*)

$\boxed{82}$ *The TI-82 has no such functions.*

Appendix: Simulating complex numbers with a TI-82

The TI-82 does not support complex numbers; however, it can be made (using matrices) to add, subtract, multiply, and divide complex numbers in rectangular format. Here are the details:

Enter two matrices in the TI-82 by reproducing the screen on the right. The "[" character is 2nd ×, and "]" is 2nd −. Enter "[A]" as MATRX 1, and "[B]" as MATRX 2 (do not use 2nd × and 2nd − for these brackets). The screen is shown just before pressing ENTER.

```
[[1,0] [0,1]]→[A]
            [[1  0]
             [0  1]]
[[0,1] [-1,0]]→[B
]■
```

The matrix [A] stands for "1", and [B] stands for "i." To enter $2 - 3i$, for example, type 2[A]-3[B] (**not** just 2-3[B]). Addition and subtraction are simply performed; the screen on the right shows the computation of the addition problem $(3+5i)+(6-2i)$. To obtain the answer—$9+3i$—simply read the first row of the resulting matrix; the first number is the real part, and the second is the imaginary part.

```
3[A]+5[B]
           [[3   5]
            [-5  3]]
Ans+(6[A]-2[B])
           [[9   3]
            [-3  9]]
■
```

Multiplication is no more complicated than addition and subtraction. Shown below are the calculator entries and outputs for some sample multiplication problems.

```
(2[A]-3[B])(3[A]
+4[B])
         [[18 -1]
          [1  18]]
■
```
$$(2 - 3i)(3 + 4i)$$
$$= 18 - i$$

```
(5[A]-4[B])(7[A]
-2[B])
         [[27 -38]
          [38  27]]
■
```
$$(5 - 4i)(7 - 2i)$$
$$= 27 - 38i$$

```
(6[A]+5[B])(6[A]
-5[B])
         [[61 0 ]
          [0  61]]
■
```
$$(6 + 5i)(6 - 5i)$$
$$= 61$$

```
(4[A]+3[B])²
         [[7   24]
          [-24 7 ]]
■
```
$$(4 + 3i)^2$$
$$= 7 + 24i$$

For division problems, do not use the ÷ key. Instead, *multiply* by the inverse (x^{-1}) of the denominator.

```
(3[A]+2[B])(5[A]
-[B])⁻¹▶Frac
       [[1/2   1/2]
        [-1/2  1/2]]
■
```
$$\frac{3 + 2i}{5 - i} = \frac{1}{2} + \frac{1}{2}i$$

```
(4[A]+2[B])(3[A]
-[B])⁻¹
       [[1   1]
        [-1  1]]
■
```
$$\frac{4 + 2i}{3 - i} = 1 + i$$

```
3[A]*[B]⁻¹
       [[0  -3]
        [3  0 ]]
■
```
$$\frac{3}{i} = -3i$$

```
(2[A]+[B])((A]-
2[B])^3)⁻¹▶Frac
     [[-4/25  -3/25]
      [3/25   -4/25]]
■
```
$$\frac{2 + i}{(1 - 2i)^3} =$$
$$-\frac{4}{25} - \frac{3}{25}i$$

Introduction

The information in this section is essentially a summary of material that can be found in the TI-86 manual. Consult that manual for more details.

While the TI-85 and TI-86 differ in some details, in most cases the instructions given in this chapter can be applied (perhaps with slight alteration) to a TI-85. The icon 85 is used to identify significant differences between the two, but some differences (e.g., a slight difference in keystrokes between the two calculators) are not noted. TI-85 users should watch for these comments.

1   Power

To power up the calculator, simply press the ON key. This should bring up the "home screen"—a flashing block cursor, and possibly the results of any previous computations that might have been done.

If the home screen does not appear, one may need to adjust the contrast (see the next section).

To turn the calculator off, press 2nd ON (note that the "second function" of ON —written in yellow type above the key—is "OFF"). The calculator will automatically shut off if no keys are pressed for several minutes.

2   Adjusting screen contrast

If the screen is too dark (all black), decrease the contrast by pressing 2nd then pressing and holding ▾. If the screen is too light, increase the contrast by pressing 2nd and then press and hold ▴.

As one adjusts the contrast, the numbers 1 through 9 will appear in the upper right corner of the screen. If the contrast setting reaches 8 or 9, or if the screen never becomes dark enough to see, the batteries should be replaced.

3   Replacing batteries

To replace the four AAA batteries, first turn the calculator off (2nd ON), then remove the back cover, remove and replace each battery, replace the back cover, then turn the calculator on again. (After replacing batteries, one may need to adjust the contrast down as described above.)

4   Basic operations

Simple computations are entered in essentially the same way they would be written. For example, to compute $2 + 17 \times 5$, press $\boxed{2}\boxed{+}\boxed{1}\boxed{7}\boxed{\times}\boxed{5}\boxed{\text{ENTER}}$ (the $\boxed{\text{ENTER}}$ key tells the calculator to act on what has been typed). Standard order of operations (including parentheses) is followed.

```
2+17*5
          87
■
```

The result of the most recently entered expression is stored in Ans, which is typed by pressing $\boxed{\text{2nd}}\boxed{(-)}$ (the word "ANS" appears in yellow above this key). For example, $\boxed{5}\boxed{+}\boxed{\text{2nd}}\boxed{(-)}\boxed{\text{ENTER}}$ will add 5 to the result of the previous computation.

```
2+17*5
          87
5+Ans
          92
■
```

After pressing $\boxed{\text{ENTER}}$, the TI-86 automatically produces Ans if the first key pressed is one which requires a number before it; the most common of these are $\boxed{+}$, $\boxed{-}$, $\boxed{\times}$, $\boxed{\div}$, $\boxed{\wedge}$, $\boxed{x^2}$, and $\boxed{\text{STO}\blacktriangleright}$. For example, $\boxed{+}\boxed{5}\boxed{\text{ENTER}}$ would accomplish the same thing as the keystrokes above (that is, it adds 5 to the previous result).

```
2+17*5
          87
5+Ans
          92
Ans+5
          97
■
```

Pressing $\boxed{\text{ENTER}}$ by itself evaluates the previously typed expression again. This can be especially useful in conjunction with Ans. The screen on the right shows the result of pressing $\boxed{\text{ENTER}}$ a second time.

```
2+17*5
          87
5+Ans
          92
Ans+5
          97
         102
■
```

Several expressions can be evaluated together by separating them with colons ($\boxed{\text{2nd}}\boxed{.}$). When $\boxed{\text{ENTER}}$ is pressed, the result of the *last* computation is displayed. The screen shown illustrates the computation $2(5 + 1)^2$.

```
3+2
           5
Ans+1:Ans²:2 Ans
          72
■
```

5   Cursors

When typing, the appearance of the cursor indicates the behavior of the next keypress. When the standard cursor (a flashing solid block, ■) is visible, the next keypress will produce its standard action—that is, the command or character printed on the key itself.

If $\boxed{\text{2nd}}\boxed{\text{DEL}}$ is pressed, the TI-86 is placed in INSERT mode and the standard cursor will appear as a flashing underscore. If the arrow keys ($\boxed{\blacktriangle}$, $\boxed{\blacktriangledown}$, $\boxed{\blacktriangleright}$, $\boxed{\blacktriangleleft}$) are used to move the cursor around within the expression, and the TI-86 is placed in INSERT mode, subsequent characters and commands will be inserted in the line at the cursor's position. When the cursor appears as a block, the TI-86 is in DELETE (or OVERWRITE) mode, and subsequent keypresses will replace the character(s) at the cursor's position. (When the cursor is at the end of the expression, this is irrelevant.)

The TI-86 will return to DELETE mode when any arrow key is pressed. It can also be returned to DELETE mode by pressing $\boxed{\text{2nd}}\boxed{\text{DEL}}$ a second time.

Pressing $\boxed{\text{2nd}}$ causes an arrow to appear in the cursor: $\boxed{\uparrow}$ (or an underscored arrow). The next keypress will produce its "second function"—the command or character printed in yellow above the key. (The cursor will then return to "standard.") If $\boxed{\text{2nd}}$ is pressed by mistake, pressing it a second time will return the cursor to standard.

Pressing ALPHA places the letter "A" in the cursor: 🅰 (or an underscored "A"). The next keypress will produce the letter or other character printed in blue above that key (if any), and the cursor will then return to standard. Pressing 2nd ALPHA puts the calculator in lowercase ALPHA mode, changing the cursor to 🄰 and producing the lowercase version of a letter. Pressing ALPHA twice (or 2nd ALPHA ALPHA) "locks" the TI-86 in ALPHA (or lowercase ALPHA) mode, so that all of the following keypresses will produce characters until ALPHA is pressed again.

6   Accessing previous entries ("deep recall")

By repeatedly pressing 2nd ENTER ("ENTRY"), previously typed expressions can be retrieved for editing and re-evaluation. Pressing 2nd ENTER once recalls the most recent entry; pressing 2nd ENTER again brings up the second most recent, etc. The number of previous entries thus displayed varies with the length of each expression (the TI-86 allocates 128 bytes to store previous expressions).

85 *The TI-85 allows access only to the last entry typed.*

7   Menus

Keys such as TABLE, GRAPH and 2nd 7 (MATRX) bring up a menu line at the bottom of the screen with a variety of options. These options can be selected by pressing one of the function keys (F1, F2, ..., F5). If the menu ends with a small triangle ("▸"), it means that more options are available in this menu, which can viewed by pressing MORE. Shown is the menu produced by pressing 2nd × (MATH).

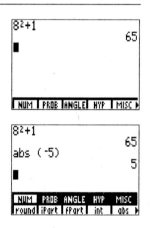

This screen shows the result of pressing F1 (the "NUM" option, which lists a variety of numerical functions). Note that the MATH menu still appears (with NUM highlighted) and the bottom line now lists the functions available in this sub-menu—including, for example, the absolute value function (abs), which is accessed by pressing F5. The command line abs (-5) was typed by pressing F5 ((-) 5) ENTER.

This manual will use (e.g.) MATH:NUM to indicate commands accessed through menus like this. Sometimes the keypresses will be included as well; for this example, it would be 2nd × F1 F5.

The various commands in these menus are too numerous to be listed here. They will be mentioned as needed in the examples.

One last comment is worthwhile, however. Some functions that may be used frequently are buried several levels deep in the menus, and may take many keystrokes to access. Worse, the location of the function might be forgotten (is it MATH:NUM or MATH:MISC?), necessitating a search through the menus. It is useful to remember three things:

- Any command can be typed one letter at a time, in either upper- or lowercase; e.g., ALPHA ALPHA LOG SIN 6 (-) will type the letters "ABS ", which has the same effect as 2nd × F1 F5.

- Any command can be found in the CATALOG menu (2nd CUSTOM F1). Since the commands appear in alphabetical order, it may take some time to locate the desired function. Pressing any letter key brings up commands starting with that letter (it is not necessary to press ALPHA first); e.g., pressing LOG brings up the list on the right, while pressing [,] shows commands starting with "P."

- Frequently used commands can be placed in the CUSTOM menu, and will then be available simply by pressing CUSTOM. To do this, scroll through the CATALOG to find the desired function, then press F3 (CUSTM) followed by one of F1–F5 to place that command in the CUSTOM menu. In the screen shown, F1 was pressed, so that pressing CUSTOM F1 will type "Solver(." The commands in the MATH:ANGLE menu (2nd × F3), used frequently for problems in this text, could be made more accessible by placing them in this menu.

8   Variables

The uppercase letters A through Z, as well as some (but not all) lowercase letters, and also sequences of letters (like "High" or "count") can be used as variables (or "memory") to store numerical values. To store a value, type the number (or an expression) followed by STO▶, then a letter or letters (note that the TI-86 automatically goes into ALPHA mode when STO▶ is pressed), then ENTER. That variable name can then be used in the same way as a number, as demonstrated at right.

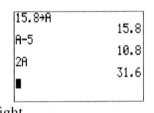

Note: The TI-86 interprets 2A as "2 times A"—the "∗" symbol is not required (this is consistent with how we interpret mathematical notation). As for order of operations, this kind of multiplication is treated the same as "∗" multiplication.

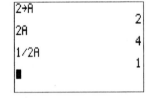

[85] *This latter comment is **not** true of the TI-85; on the TI-85, implied multiplication (such as 2A) is done before other multiplication and division, and even before some other operations, like the square root function ∫. Therefore, for example, the expression 1/2A is evaluated as 1/4 on the TI-85 (assuming that A is 2).*

9   Setting the modes

By pressing 2nd MORE (MODE), one can change many aspects of how the calculator behaves. For most of the examples in this manual, the "default" settings should be used; that is, the MODE screen should be as shown on the right. Each of the options is described below; consult the TI-86 manual for more details. Changes in the settings are made using the arrows keys and ENTER.

The Normal Sci Eng setting specifies how numbers should be displayed. The screen on the right shows the number 12345 displayed in Normal mode (which displays numbers in the range $\pm 999,999,999,999$ with no exponents), Sci mode (which displays all numbers in scientific notation), and Eng mode (which uses only exponents that are multiples of 3). Note: "E" is short for "times 10 to the power," so $1.2345\text{E}4 = 1.2345 \times 10^4 = 1.2345 \times 10000 = 12345$.

The Float 012345678901 setting specifies how many places after the decimal should be displayed (the 0 and 1 at the end mean 10 and 11 decimal places). The default, Float, means that the TI-86 should display all non-zero digits (up to a maximum of 12).

Radian Degree indicates whether angle measurements should be assumed to be in radians or degrees. (A right angle measures $\frac{\pi}{2}$ radians, which is equivalent to $90°$.) This text does not refer to angle measurement.

RectC PolarC specifies whether complex numbers should be displayed in rectangular or polar format. This text uses the $a + bi$ format (similar to the TI-86's rectangular format) for complex numbers. More information about complex numbers can be found beginning on page 44 (Section 1.3, Example 1) of this manual.

Func Pol Param DifEq specifies whether formulas to be graphed are functions (y as a function of x), polar equations (r as a function of θ), parametric equations (x and y as functions of t), or differential equations ($Q'(t)$ as a function of Q and t). The text accompanying this manual uses only the first of these modes.

The RectV CylV SphereV setting indicates the default display format for vectors (not used in this text).

The other two mode settings deal with issues that are beyond the scope of the textbook, and are not discussed here.

A group of settings related to the graph screen are found by pressing GRAPH MORE F3 (GRAPH:FORMT). The default settings are shown in the screen on the right, and are generally the best choices for most examples in this book (although the last setting could go either way).

RectGC PolarGC specifies whether graph coordinates should be displayed in rectangular (x, y) or polar (r, θ) format. Note that this choice is independent of the Func Pol Param DifEq mode setting.

The CoordOn CoordOff setting determines whether or not graph coordinates should be displayed.

When plotting a graph, the DrawLine DrawDot setting tells the TI-86 whether or not to connect the individually plotted points. SeqG SimulG specifies whether individual expressions should be graphed one at a time (sequentially), or all at once (simultaneously).

GridOff GridOn specifies whether or not to display a grid of dots on the graph screen, while AxesOn AxesOff and LabelOff LabelOn do the same thing for the axes and labels (y and x) on the axes.

10   Setting the graph window

Pressing GRAPH F2 brings up the WINDOW settings. The exact contents of the WINDOW menu vary depending on whether the calculator is in function, polar, parametric, or DifEq mode; below are four examples showing this menu in each of these modes.

Function mode

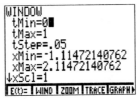
Polar mode

Parametric mode

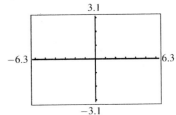
DifEq mode

85 *On the TI-85, the* WINDOW *settings are called the* RANGE *settings.*

All these menus include the values xMin, xMax, xScl, yMin, yMax, and yScl. When GRAPH F5 (GRAPH) is pressed, the TI-86 will show a portion of the Cartesian (x-y) plane determined by these values. In function mode, this menu also includes xRes, the behavior of which is described in section 12 of this manual (page 38). The other settings in this screen allow specification of the smallest, largest, and step values of θ (for polar mode) or t (for parametric mode), or initial conditions for the differential equation.

With settings as in the "Function mode" screen shown above, the TI-86 would display the screen at right: x values from -6.3 to 6.3 (that is, from xMin to xMax), and y values between -3.1 to 3.1 (yMin to yMax). Since xScl = yScl = 1, the TI-86 places tick marks on both axes every 1 unit; thus the x-axis ticks are at $-6, -5, \ldots, 5$, and 6, and the y-axis ticks fall on the integers from -3 to 3. This window is called the "decimal" window, and is most quickly set by pressing GRAPH F3 (ZOOM) MORE F4 (ZDECM).

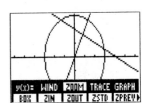

Note: If the graph screen has a menu on the bottom (like that shown on the right), possibly obscuring some important part of the graph, it can be removed by pressing CLEAR. The menu can be restored later by pressing EXIT.

Below are four more sets of window settings, and the graph screens they produce. Note that the first graph on the left has tick marks every 10 units on both axes. The second window is called the "standard" viewing window, and is most quickly set by pressing [GRAPH][F3] (ZOOM) [F4] (ZSTD). The setting $yScl = 0$ in the final graph means that no tick marks are placed on the y-axis.

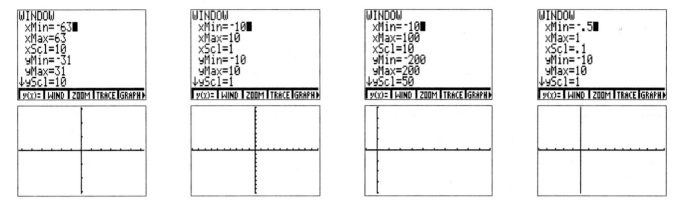

11   The graph screen

The TI-86 screen is made up of an array of rectangular dots (pixels) with 63 rows and 127 columns. All the pixels in the leftmost column have x-coordinate xMin, while those in the rightmost column have x-coordinate xMax. The x-coordinate changes steadily across the screen from left to right, which means that the coordinate for the nth column (counting the leftmost column as column 0) must be $\text{xMin} + n\Delta x$, where $\Delta x = (\text{xMax} - \text{xMin})/126$. Similarly, the nth row of the screen (counting up from the bottom row, which is row 0) has y-coordinate $\text{yMin} + n\Delta y$, where $\Delta y = (\text{yMax} - \text{yMin})/62$.

It is not necessary to memorize the formulas for Δx and Δy. Should they be needed, they can be determined by pressing [GRAPH][F5] and then the arrow keys. When pressing [▶] or [◀] successively, the displayed x-coordinate changes by Δx; meanwhile, when pressing [▲] or [▼], the y-coordinate changes by Δy. Alternatively, the values can be found by typing "Δx" and "Δy" on the home screen; this 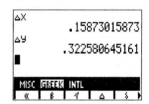 is most easily done by pressing [2nd][0][F2][F4] to access the CHAR:GREEK menu and type the "Δ" character, then typing lowercase x or y. This produces results like those shown on the right; the CHAR:GREEK menu remains on the bottom of the screen.

In the decimal window $\text{xMin} = -6.3$, $\text{xMax} = 6.3$, $\text{yMin} = -3.1$, $\text{yMax} = 3.1$, note that $\Delta x = 0.1$ and $\Delta y = 0.1$. Thus, the individual pixels on the screen represent x-coordinates $-6.3, -6.2, -6.1, \ldots, 6.1, 6.2, 6.3$ and y-coordinates $-3.1, -3, -2.9, \ldots, 2.9, 3, 3.1$. This is where the decimal window gets its name.

It happens that the pixels on the TI-86 screen are about 1.2 times taller than they are wide, so if $\Delta y/\Delta x$ is approximately 1.2 (the exact value is $1.19565\ldots$), the window will be a "square" window (meaning that the scales on the x- and y-axes are equal). For example, the decimal window (with $\Delta y/\Delta x = 1$) is not square, so that one unit on the x-axis is not the same length as one unit on the y-axis. (Specifically, one y-axis unit is about 20% longer than one x-axis unit.)

Any window can be made square be pressing [GRAPH][F3] (ZOOM) [MORE][F2] (ZSQR). To see the effect of a square window, observe the two pairs of graphs below. In each pair, the first graph is on the standard window, and the second is on a square window (after choosing ZOOM:ZSQR). The first pair shows the lines $y = 2x - 3$ and $y = 3 - \frac{1}{2}x$; note that on the square window, these lines look perpendicular (as they should). The second pair shows a circle centered at the origin with a radius of 8. On the standard window, this looks like an oval since the screen is wider than it is tall. (The reason for the gaps in the circle will be addressed in the next section.)

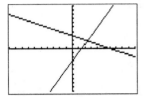 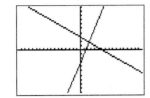 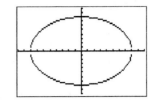 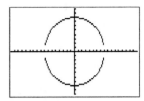

12   Graphing a function

This introductory section only addresses creating graphs in function mode. This textbook does not include examples of parametric and polar graphs; however, procedures for creating such graphs are very similar.

To see the graph of $y = 2x - 3$, begin by entering the formula into the calculator. This is done by pressing [GRAPH][F1] to access the "y equals" screen of the calculator. Enter the formula as y1 (or any other yn); note that the letter x can be typed by pressing [F1] or [x-VAR] (as well as [2nd][ALPHA][+]). If another y variable has a formula, position the cursor on that line and press either [F4] (DELf—to delete the function) or [F5] (SELCT). The latter has the effect of toggling the "highlighting" for the equals sign "=" for that line (an "unhighlighted" equals sign tells the TI-86 not to graph that formula). In the screen on the right, only y1 will be graphed.

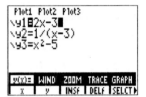

The next step is to choose a viewing window. See the previous section for more details on this. This example uses the standard window ([GRAPH][F3][F4]).

If the graph has not been displayed, press [GRAPH][F5], and the line should be drawn. In order to produce this graph, the TI-86 considers 127 values of x, ranging from xMin to xMax in steps of Δx (assuming that xRes = 1; see below for other possibilities). For each value of x, it computes the corresponding value of y, then plots that point (x, y) and (if the calculator is in Connected [DrawLine] mode) draws a line between this point and the previous one.

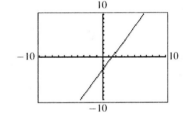

If xRes is set to 2, the TI-86 will only compute y for every other x value; that is, it uses a step size of $2\Delta x$. Similarly, if xRes is 3, the step size will be $3\Delta x$, and so on. Setting xRes higher causes graphs to appear faster (since fewer points are plotted), but for some functions, the graph may look "choppy" if xRes is too large, since detail is sacrificed for speed.

Note: If the line does not appear, or the TI-86 reports an error, double-check all the previous steps. Also, check the mode settings (discussed in section 9, page 34).

Once the graph is visible, the window can be changed using [F2] (WINDOW) or [F3] (ZOOM). Pressing [F4] (TRACE) brings up the "trace cursor," and displays the x- and y-coordinates for various points on the line as the [◄] and [►] keys are pressed. Tracing beyond the left or right columns causes the TI-86 to adjust the values of xMin and xMax and redraw the graph.

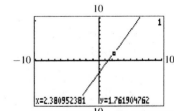

To graph the function

$$y = \frac{1}{x-3},$$

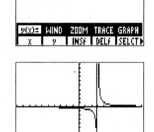

enter that formula into the "y equals" screen (note the use of parentheses). As before, this example uses the standard viewing window.

For this function, the TI-86 produces the graph shown on the right. This illustrates one of the pitfalls of the connect-the-dots method used by the calculator: The nearly-vertical line segment drawn at $x = 3$ *should not be there*, but it is drawn because the calculator connects the points

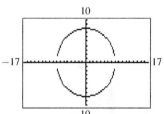

$$x = 2.85714, \ y = -6.99999 \ \text{and} \ x = 3.01587, \ y = 62.99999.$$

Calculator users must learn to recognize these flaws in calculator-produced graphs.

The graph of a circle centered at the origin with radius 8 (shown on the square window ZOOM:ZSTD - ZOOM:ZSQR) shows another problem that arises from connecting the dots. When $x = -8.093841$, y is undefined, so no point is plotted (that is, there is no point on this circle that has x-coordinate less than -8, or greater than 8). The next point plotted on the upper half of the circle is $x = -7.824046$ and $y = 1.668619$; since no point had been plotted for the previous x-coordinate, this is not connected to anything, so there appears to be a gap between the circle and the x-axis. The calculator is not "smart" enough to know that the graph should extend from -8 to 8.

One additional feature of graphing with the TI-86 is that each function can have a "style" assigned to its graph. The symbol to the left of y1, y2, etc. indicates this style, which can be changed by choosing GRAPH:y(x)=:STYLE to cycle through the options. These options are shown on the right (with brief descriptive names); more information can be found in the examples, and complete details are in the TI-86 manual. [85] *The TI-85 does not include graph-style features.*

---

13   Adding programs to the TI-86

The TI-86's capabilities can be extended by downloading or entering programs into the calculator's memory. Instructions for writing a program are beyond the scope of this manual, but programs written by others and downloaded from the Internet (or obtained as printouts) can be transferred to the calculator in one of three ways:

1. If one TI-86 already has a program, it can be transferred to another using the calculator-to-calculator link cable. To do this, first make sure the cable is firmly inserted in both calculators. On the sending calculator, press [2nd][x-VAR] (LINK), then [F1]:[F2] (SEND:PRGM), and then select (by using the [▲] and [▼] keys and [F2]) the program(s) to be transferred. *Before* pressing [F1] (SEND) on the sending calculator, prepare the receiving calculator by pressing [2nd][x-VAR][F2], and *then* press [F1] on the sending calculator.

2. If a computer with the TI-Graph Link is available, and the program file is on that computer (e.g., after having been downloaded from the Internet), the program can be transferred to the calculator using the TI Connect (or TI Graph Link) software. This transfer is done in a manner similar to the calculator-to-calculator transfer described above; specific instructions can be found in the documentation that

accompanies the software. (They are not given here because of slight differences between platforms and software versions.)

3. View a listing of the program and type it in manually. (**Note:** Even if the TI-Graph Link cable is not available, the software can be used to view program listings on a computer.) While this is the most tedious method, studying programs written by others can be a good way to learn programming. To enter a program, start by choosing [PRGM][F2] (EDIT), then type a name for the new program (up to eight letters, like "QuadForm" or "Midpoint")—note that the TI-86 is automatically put into ALPHA mode. Then type each command in the program, and press [2nd][EXIT] (QUIT) to return to the home screen when finished.

To run the program, make sure there is nothing on the current line of the home screen, then press [PRGM][F1], select the program using one of the keys [F1]–[F5] and [MORE] (a sample screen is shown; only the first four to six characters of each program name are shown), and press [ENTER]. If the program was entered manually (option 3 above), errors may be reported; in that case, choose GOTO, correct the mistake and try again.

Programs can be found at many places on the Internet, including:

- http://www.bluffton.edu/~nesterd—the Web site of the author of this manual;

- http://tifaq.calc.org—a "Frequently Asked Questions" page maintained by Ray Kremer; and

- http://www.ticalc.org.

Examples

Here are the details for using the TI-86 for several of the examples from the textbook. Also given are the keystrokes necessary to produce some of the commands shown in the text's examples. In some cases, some suggestions are made for using the calculator more efficiently.

Throughout this section, it is assumed that the textbook is available for reference. The problems from the text are not restated here, and there are frequent references to the calculator screens shown in the text.

Section R.2   Example 2   (page 10)   Evaluating Exponential Expressions

---

This example discusses how to evaluate 4^3, $(-6)^2$, -6^2, $4 \cdot 3^2$, and $(4 \cdot 3)^2$.

To evaluate 4^3 with a TI-86, type $\boxed{4}\boxed{\wedge}\boxed{3}\boxed{\text{ENTER}}$; the TI-86 does not have a superscripted 3 for cubing. Only squaring and reciprocating—that is, the powers 2 and -1—can be displayed as superscripts (using $\boxed{x^2}$ and $\boxed{\text{2nd}}\boxed{\text{EE}}$, respectively). All other powers must use the $\boxed{\wedge}$ key. All other computations in this example are straightforward with the TI-86, except for this *caveat*: For $(-6)^2$ and -6^2, one

```
4^3
              64
(-6)²
              36
-6²
             -36
(-6)²█
```

must use the $\boxed{(-)}$ key rather than the $\boxed{-}$ key to enter "negative 6." In the screen shown, the second line was typed as $\boxed{(}\boxed{(-)}\boxed{6}\boxed{)}\boxed{x^2}$, while the fourth line was $\boxed{(}\boxed{-}\boxed{6}\boxed{)}\boxed{x^2}$. When $\boxed{\text{ENTER}}$ is pressed, the fourth line produces a syntax error—the calculator's way of saying that the line makes no sense.

Section R.2   Example 3   (page 11)   Using Order of Operations

---

Using the calculator for parts (a) and (b) of this example is straightforward; the expressions are entered as printed. For (c) and (d), we have the expressions

$$\frac{4 + 3^2}{6 - 5 \cdot 3} \quad \text{and} \quad \frac{-(-3)^3 + (-5)}{2(-8) - 5(3)}.$$

Entering these on the TI-86 requires us to add some additional parentheses to make sure the correct order of operations is followed.

$\boxed{85}$ *TI-85 users: See the note about "implied multiplication" on page 34.*

Proper entry of the expression in (d) is shown on the right, in two different forms. The second illustrates some time-saving that can be done in keying in such expressions: The parentheses around -5 in the numerator were omitted, and the denominator $2(-8) - 5(3)$ was typed in as $\boxed{2}\boxed{\times}\boxed{(-)}\boxed{8}\boxed{-}\boxed{5}\boxed{\times}\boxed{3}$—the parentheses were exchanged for the multiplication symbol. It would be fine to type the

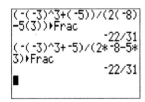

```
(-(-3)^3+(-5))/(2(-8)
-5(3))▶Frac
                -22/31
(-(-3)^3+-5)/(2*-8-5*
3)▶Frac
                -22/31
█
```

expression unchanged, but the extra parentheses make it somewhat harder to read (and also mean more opportunities to make a mistake).

The screen shown above also illustrates the command ▶Frac at the end of the line. This command (which simply means "display the result of this computation as a fraction, if possible") is available in the MATH:MISC menu ($\boxed{\text{2nd}}\boxed{\times}\boxed{\text{F5}}\boxed{\text{MORE}}\boxed{\text{F1}}$). This is a useful enough command that one may wish to put it in the CUSTOM menu (see page 34).

Section R.2   Example 7   (page 15)   Evaluating Absolute Value

Section R.2   Example 9   (page 16)   Evaluating Absolute Value Expressions

The absolute value function abs is found in the MATH:NUM menu (2nd × F1 F5).
See the discussion of typing commands on page 33; the screen on the right illustrates that the TI-86 does not pay attention to upper- and lowercase.

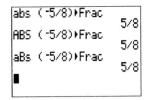

Section R.6   Example 4   (page 55)   Using the Definition of $a^{1/n}$

In evaluating these fractional exponents with a calculator, some time can be saved
by entering fractions as decimals. The screen on the right performs the computations for (a), (b) and (c) with fewer keystrokes. (Of course, as noted in Section R.7
of the text, $36^{1/2} = \sqrt{36}$, which requires even fewer keystrokes.)

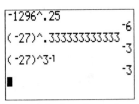

Note for (d), the parentheses around -1296 cannot be omitted: The text observes
that this is not a real number, but the calculator will display a real result if the
parentheses are left out. (See also the next example.) For (e), the decimal equivalent of $1/3$ is $0.\overline{3}$, which can be entered as a sufficiently long string of 3s (at least
12). The last line shows a better way to do this; the TI-86's order of operations
is such that 3^{-1} is evaluated before the other exponentiation. This method would also work for the other
computations; e.g., (a) could be entered 3 6 ^ 2 2nd EE .

Section R.6   Example 5   (page 56)   Using the Definition of $a^{m/n}$

For (f), attempting to evaluate $(-4)^{5/2}$ will produce the result shown on the right.
This is the TI-86's way of displaying complex numbers (which are discussed in
the text beginning in Section 1.3); the result $(0, 32)$ would be written as $0 + 32i$
in the format used by the text.

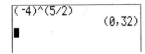

Section R.7   Example 1   (page 63)   Evaluating Roots

Aside from the square root function $\int$ (2nd x^2), the TI-86 has an "x-root" function ($^x\int$), found in the
MATH:MISC menu (2nd × F5 MORE F4). For example, "3$^x\int$ 1000" would compute $\sqrt[3]{1000}$ for part (d) . It is
worth noting that fourth roots can be typed in more efficiently using two square roots, since $\sqrt[4]{a} = a^{1/4} =$
$(a^{1/2})^{1/2} = \sqrt{a^{1/2}} = \sqrt{\sqrt{a}}$.

Section 1.1   Example 1   (page 85)   Solving a Linear Equation

The TI-86 offers several approaches to solving (or confirming a solution to) an
equation. Aside from graphical approaches to solving equations (which are discussed later, on page 50), here are some non-graphical (numerical) approaches:
As illustrated on the right, the TI-86's `Solver` function (found in the CATALOG)
attempts to find a value of x that makes the given expression equal to 0, given a
guess (5, in this case). The solution is stored in the variable x (or whatever variable
is used in the equation), but as the screen shows, this solution is not automatically
displayed. A colon can be used to place the two lines shown on a single line.

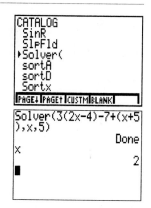

For the equations in this section, the "guess" value does not affect the result, but for equations in later
sections, different guesses might produce different results. Full details on how to use this function can be
found in the TI-86 manual.

Since it is difficult to access the `Solver` command, it might be a good idea to place `Solver` in the CUSTOM
menu (see page 34), or to use the "interactive solver" built-in to the TI-86. This feature is found by pressing
2nd GRAPH.

Solver begins with the first screen below (a), requesting an equation, which must include at least one
variable. (The "eqn:" line might not be blank, since Solver remembers the most recent equation entered
into it.) The menu at the bottom of the screen lists any user-defined variables in the calculator's memory.
The expression for this example is shown in screen (b); use ALPHA STO▸ to type the equals sign. Pressing
ENTER or ▾ brings up screen (c). Type the guess (this example uses 5), then press F5 (SOLVE) and the
TI-86 seeks a solution near 5, which is reported in screen (d). (Make sure the cursor is on the "x=" line
when F5 is pressed.)

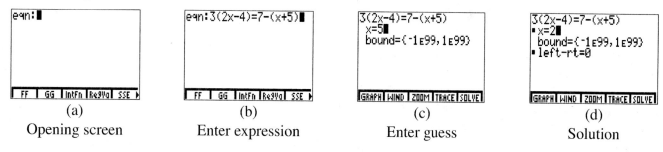

(a)	(b)	(c)	(d)
Opening screen	Enter expression	Enter guess	Solution

From screen (c) or (d), pressing ▲ returns to screen (b), so that the equation can be edited.

The Solver can also be used with equations containing more than one variable; simply provide values for
all but one variable, then place the cursor on the line containing the variable for which a value is needed
and press F2.

Note: In this example, we learned how the TI-86 can be used to support an analytic solution. But the
TI-86 and any other graphing calculator also can be used for solving problems when an analytic solution
is **not** possible — that is, when one cannot solve an equation "algebraically." This is often the case in many
"real-life" applications, and is one of the best arguments for the use of graphing calculators.

Section 1.3   Example 1   (page 104)   Writing $\sqrt{-a}$ as $i\sqrt{a}$

Section 1.3   Example 2   (page 105)   Finding Products and Quotients Involving Negative Radicands

Section 1.3   Example 3   (page 105)   Simplifying a Quotient Involving a Negative Radicand

The TI-86 does these computations almost as shown in the text, except that it reports the results differently. The screen on the right illustrates the output from the computations for parts (b) and (d) of Example 2. The TI-86 uses the format (a, b) for the complex number $a + bi$; whenever the TI-86 does a computation involving square roots of negative numbers, it reports the results in this format *even if the result is real*. Thus $(-7.74596669242, 0)$ represents the real number $-7.74596669242 \approx -2\sqrt{15}$, and $(0, 1.41421356237)$ is $i\sqrt{2}$.

While the calculator always reports complex *results* in this format*, it can recognize *input* typed in using the text's format by first defining a variable i as the complex number $(0, 1) = 0 + 1i$. Shown on the right is a computation using this definition. One could also use I (uppercase), which can be typed with one less keystroke, but i will be used in these examples.

*Results can also be reported in "polar format" by changing the "RectC PolarC" mode setting, but this format is even further removed from the $a + bi$ format used in the text.

Section 1.3   Example 4   (page 106)   Adding and Subtracting Complex Numbers

Section 1.3   Example 5   (page 107)   Multiplying Complex Numbers

Section 1.3   Example 6   (page 107)   Simplifying Powers of *i*

Section 1.3   Example 7   (page 108)   Dividing Complex Numbers

Using the variable i defined to be $(0, 1)$—as was described in the previous example—computations like these can be done in a manner similar to those illustrated in the text's calculator screens. The screens on the right show (in the TI-86's format) the same results as those in the text accompanying Examples 5 and 7.

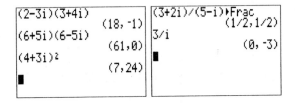

Section 1.4   Example 5   (page 115)   Using the Quadratic Formula (Real Solutions)

Section 1.4   Example 6   (page 115)   Using the Quadratic Formula (Nonreal Complex Solutions)

Section 1.4   Example 7   (page 116)   Solving a Cubic Equation

The TI-86 can solve quadratic and cubic equations (as well as higher-degree polynomial equations) using its built-in polynomial solver, accessed through [2nd] [PRGM] (POLY). This first prompts the user for the "order" (degree) of the polynomial, meaning the highest power of x. For the quadratic equations of Examples 5 and 6, this should be 2; for Example 7, the order is 3.

Pressing [ENTER] then brings up the screen on the right, requesting the coefficients of the equation. Note that the top line of the screen contains a reminder that the expression must be equal to 0. The menu at the bottom indicates that [F1] will clear the coefficients, while [F5] solves the equation.

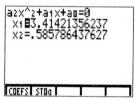

Pressing [F5] reports the solutions in much the same way as the program shown in the text (although it does not specify that the roots are real). [F1] allows the user to change the coefficient values (that is, it goes back to the previous screen), and [F2] provides a way to store the coefficients in a variable (as a list).

For Example 6, the TI-86 reports the solutions (using its complex number format) as approximately $0.25 \pm 1.3919i$ —the decimal equivalents of the answers given in the text. Note in the results for Example 7, the real solution -2 is displayed as $(-2, 0) = -2 + 0i$; if any of the solutions are complex, the TI-86 displays all of them in its complex format.

Section 1.5   Example 1   (page 122)   Solving a Problem Involving the Volume of a Box

A *table* can be a useful tool to solve equations like this. To use the table features of the TI-86, begin by entering the formula ($y = 15x^2 - 200x + 500$) on the GRAPH:y(x)= screen, as one would to create a graph. (The highlighted equals signs determine which formulas will be displayed in the table, just as they do for graphs.)

Next, press [TABLE][F2] to access the TABLE SETUP screen. The table will display y values for given values of x. The TblStart value sets the lowest value of x, while ΔTbl determines the "step size" for successive values of x. These two values are only used if the Indpnt option is set to Auto—this means, "automatically generate the values of the independent variable (x)." The effect of setting this option to Ask is illustrated at the end of this example.

When the TABLE SETUP options are set satisfactorily, press F1 (TABLE) to produce the table. Shown (above, right) is the table generated based on the settings in the above screen; the fact that y is negative for $x = 8$ and $x = 9$ supports the restriction $x > 10$ from Step 3 of the example. By pressing ⌄ repeatedly, the x values are increased, and the y values updated. (Pressing ⌃ decreases the x values, but clearly that is not appropriate for this problem.) After pressing ⌄ ten times, the table looks like the second screen, which shows that y1 equals 1435 when x equals 17.

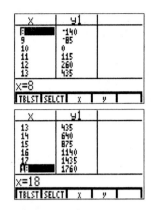

Extension: Suppose the specifications called for the box to be 1500 cubic inches (instead of 1435). We see from the table above that x must be between 17 and 18 inches. The TI-86's table capabilities provide a convenient way to "zoom in" on the value of x (which could also be found with the Solver, or by other methods).

Since $17 < x < 18$, set TblStart to 17 and ΔTbl to 0.1. This produces the table shown on the right, from which we see that $17.2 < x < 17.3$. Now set TblStart to 17.2 and ΔTbl to 0.01, which reveals that x is between 17.2 and 17.21. This process can be continued to achieve any desired degree of accuracy.

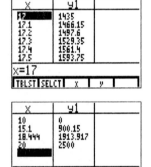

Finally, the screen on the right shows the effect of setting Indpnt to Ask. Initially, the table is blank, but as values of x are entered, the y values are computed. Up to six x values can be entered; if more are desired, the DEL key can be used to make room.

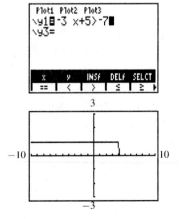

85 *The TI-85 does not have a built-in table generator, but programs are available that can be used to simulate this feature. See section 13 of the introduction (page 39) for information about installing and running programs.*

| Section 1.7 | Example 1 | (page 146) | Solving a Linear Inequality |
| Section 1.7 | Example 2 | (page 147) | Solving a Linear Inequality |

Shown is one way to visualize the solution to inequalities like these. The inequality symbols $>$, $<$, $\geq$, $\leq$ are found in the TEST menu (2nd 2nd ×). The TI-86 responds with 1 when a statement is true and 0 for false statements. This is why the graph of y1 appears as it does: For values of x less than 4, the inequality is true (and so y1 equals 1), and for $x \geq 4$, the inequality is false (and so y1 equals 0). Note that this picture does *not* help one determine what happens when $x = 4$.

Section 1.7   Example 3   (page 148)   Solving a Three-Part Inequality

To see a "picture" of this inequality like the one shown for the previous two examples requires a little additional work. Setting `y1=-2<5+3x<20` will *not* work correctly. Instead, one must enter either

   `y1=(-2<5+3x)(5+3x<20)`   or   `y1=(-2<5+3x) and (5+3x<20)`

which will produce the desired results: A function that equals 1 between $-\frac{7}{3}$ and 5, is equals 0 elsewhere. ("`and`" is most easily entered either by selecting it from the CATALOG, or by simply typing it letter-by-letter; note the spaces before and after, typed by pressing 2nd (-)).

Section 1.7   Example 5   (page 149)   Solving a Quadratic Inequality
Section 1.7   Example 6   (page 150)   Solving a Quadratic Inequality
Section 1.7   Example 8   (page 152)   Solving a Rational Inequality
Section 1.7   Example 9   (page 153)   Solving a Rational Inequality

The methods described above can also be applied to these inequalities.

Extension: Can the calculator be used to solve the inequality $x^2 + 6x + 9 > 0$? Yes—if it is used carefully. Shown is the graph of the expression $x^2 + 6x + 9 > 0$. This suggests that the inequality is true for all x; that is, the solution set is $(-\infty, \infty)$.

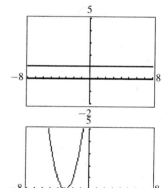

This conclusion is **not correct**: The graph of $y = x^2 + 6x + 9$ (without "> 0") touches the x-axis at $x = -3$, which means that this must be excluded from the solution set, since "> 0" means the graph must be above (not on) the x-axis. The correct solution set is $(-\infty, -3) \cup (-3, \infty)$.

This example should serve as a warning: Sometimes, the TI-86 will mislead you. When possible, try to confirm the calculator's answers in some way.

Section 1.8   Example 1   (page 159)   Solving Absolute Value Equations
Section 1.8   Example 2   (page 160)   Solving Absolute Value Inequalities

The "solver" features described on page 43 of this manual can be used for these equations, too. Recall that `abs` is entered as 2nd × F1 F5.

Section 2.1   Example 5   (page 186)   Using the Midpoint Formula

The TI-86 can do midpoint computations nicely by putting coordinates in parentheses, as shown on the right. (The TI-86 interprets an ordered pair such as (8,-4) as the complex number $8 - 4i$, but since adding two complex numbers means adding their corresponding parts, the computations are done in the correct way to find the midpoint.)

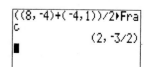

Section 2.2 Example 2 (page 194) — Graphing Circles

The text suggests graphing y1=4+√(36-(x+3)²) and y2=4-√(36-(x+3)²). Here are three options to speed up entering these formulas:

- After typing the formula in y1, move the cursor to y2 and press [2nd][STO▸] (RCL), then type y1 [ENTER] (note that [F2] will produce the lowercase y). This will "recall" the formula of y1, placing that formula in y2. Now edit this formula, changing the first "+" to a "−."

- After typing the formula in y1, enter y2=8-y1. This produces the desired results, since $8 - y1 = 8 - (4 + \sqrt{36 - (x + 3)^2}) = 8 - 4 - \sqrt{36 - (x + 3)^2} = 4 - \sqrt{36 - (x + 3)^2}$.

- Enter the single formula y1=4+{-1,1}√(36-(x+3)²). (The curly braces { and } are [F1] and [F2] in the LIST menu, [2nd][−]). When a list (like {-1,1}) appears in a formula, it tells the TI-86 to graph this formula several times, using each value in the list.

The window chosen in the text is a square window (see Section 11 of the introduction of this chapter), so that the graph looks like a circle. (On a non-square window, the graph would look like an ellipse—that is, a distorted circle.) Note, however, that on the TI-86, this window is not square. Other square windows would also produce a "true" circle, but some will leave gaps similar to those circles shown in sections 11 and 12 of the introduction (pages 37–38). Shown is the same circle on the square window produced by "squaring up" the standard window—GRAPH:ZOOM:ZSTD - ZOOM:ZSQR.

Section 2.3 Example 6 (page 208) — Using Function Notation
Section 2.3 Example 7 (page 209) — Using Function Notation

The TI-86 will do *some* computations in function notation. Simply enter the function in y1 (or any other function). Return to the home screen and enter y(2) to evaluate that function with the input 2. The result shown on the screen agrees with (a) in the text.

The TI-86 is less helpful if asked to evaluate expressions like those given in Example 6(b) or Example 7. For example, "y1(q)" produces an error message or an unpredictable result (depending on how the variable q is defined), rather than the desired result $-q^2 + 5q - 3$.

[85] *The TI-85 does not support function notation. Something like function notation can be achieved, either by typing (e.g.) 3→x:y1 (which displays the value of y1 after storing 3 in the variable x), or by using the* evalF *(evaluate function) command, found in the* CALC *menu (*[2nd][÷]*), for which (e.g.)* evalF(y1,x,3) *will evaluate y1 with x equal to 3. Entering* y1(3) *does not produce the desired results; it simply multiplies the value of y1 (using the current value of x) by 3.*

Section 2.4 Example 3 (page 219) — Graphing a Vertical Line

The text notes that such a line is not a function; therefore, it cannot be graphed by entering the equation on the GRAPH:y(x)= screen. However, the TI-86 does provide a fairly straightforward method of graphing vertical lines using the Vert command.

From the home screen, type the Vert command one letter at a time, or using the function catalog (see page 34), then type the x-coordinate of the line. When ENTER is pressed, the line will be drawn.

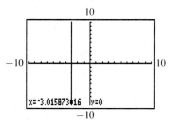

Alternatively, the vertical line can be drawn "interactively" if the VERT command is accessed through the GRAPH:DRAW menu (GRAPH MORE F2 F3). This allows the user to move the cursor (and the vertical line) to the desired position using the ◄ and ► keys. Pressing ENTER places the line at that location, then allows more lines to be drawn. (Press GRAPH to finish drawing vertical lines.) Note that since $x = -3$ does not correspond to a column of pixels on the screen (see page 37), the vertical line is actually located at $x \approx -3.0159$, but the appearance on the screen is the same as if it were at $x = -3$.

Section 2.5   Example 6   (page 236)   Finding Equations of Parallel and Perpendicular Lines

In the discussion following this example (on page 237), the text notes that the lines from part (b) do not appear to be perpendicular unless they are plotted on a square window. See section 11 of the introduction of this chapter (page 37) for more information about square windows.

Section 2.5   Example 7   (page 238)   Finding an Equation of a Line That Models Data
Section 2.5   Example 8   (page 239)   Finding a Linear Equation That Models Data

Aside from the methods described in these two examples, the text refers to the technique of linear regression on page 241. Figure 51 illustrates the steps on a TI-83; here is a more detailed description of the process on a TI-86 (using the data of Example 8).

Given a set of data pairs (x, y), the TI-86 can find various formulas (including linear and quadratic, as well as more complex formulas) that approximate the relationship between x and y. These formulas are called "regression formulas."

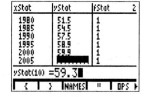

The first step in determining the regression formula is to enter the data into the TI-86. This is done by pressing 2nd + F2 (STAT:EDIT), which should produce the screen on the right. If xStat, yStat, and fStat are not on the STAT:EDIT screen, and pressing ◄ and ► does not reveal them, they can be restored to this screen by entering the command SetLEdit ("set up the list editor") on the home screen. This command is available by pressing 2nd − F5 MORE MORE MORE F3.

In the STAT:EDIT screen, enter the year values into the first column (xStat) and the percentage of women into the second column (yStat). The third column (fStat) should contain ten "1"s; this tells the TI-86 that each of these (x, y) pairs occurs only once in the data list. If any column already contains data, the DEL key can be used to delete numbers one at a time, or—to delete the whole column at once—press the ▲ key until the cursor is at the top of the column (on xStat, yStat, or fStat) and press CLEAR ENTER. Make sure that all three columns contain the same number of entries.

To do the computation, press [EXIT], then [2nd][+][F1] (STAT:CALC). The bottom of the screen now lists the various options for the type of calculation to be done. We want to do a linear regression, which is [F3] (LinR).

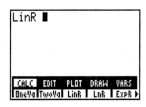

Pressing [ENTER] displays the results of the regression (shown on the right). These numbers agree (after rounding) with those in the text, except that the roles of a and b have been switched; these numbers are for a model of the form y=a+bx, rather than y=ax+b.

[85] *The TI-85 can also perform regression computations, but the steps in the process are rather different from those described here. Consult your TI-85 owner's manual for these details.*

Section 2.5   Example 9   (page 241)   Solving an Equation with a Graphing Calculator

The text illustrates a graphical process for solving the equation $-2x - 4(2-x) = 3x + 4$; in this discussion, we will show two graphical methods for solving the equation $\frac{1}{2}x - 6 = \frac{3}{4}x - 9$. (The answer is $x = 12$.)

The first method is the **intersection** method. To begin, set up the TI-86 to graph the left side of the equation as y1, and the right side as y2. [85] *Putting the fractions in parentheses ensures no mistakes with order of operations. This is not crucial for the TI-86, but is a good practice because some other models (including the TI-85) give priority to implied multiplication. See section 8 of the introduction, page 34.*

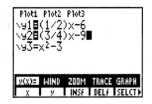

We are looking for an x value that will make the left and right sides of this equation equal to each other, which corresponds to the x-coordinate of the point of intersection of these two graphs.

After selecting a viewing window which shows the point of intersection—in this example, $[-15, 15] \times [-10, 10]$ is a convenient choice—the TI-86 can automatically locate this point using GRAPH:MATH:ISECT ([GRAPH][MORE][F1][MORE][F3]). Use [▲], [▼] and [ENTER] to specify which two functions to use (in this case, the only two being displayed), and then use [◄] or [►] to specify a guess. After pressing [ENTER], the TI-86 will try to find an intersection of the two graphs. The screens below illustrate these steps.

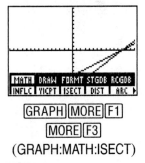

[GRAPH][MORE][F1]
[MORE][F3]
(GRAPH:MATH:ISECT)

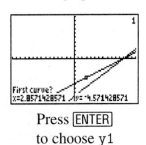

Press [ENTER]
to choose y1

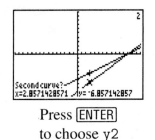

Press [ENTER]
to choose y2

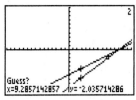

Move cursor to
specify guess and
then press [ENTER].

The final result of this process is the screen shown on the right. The *x*-coordinate of this point of intersection is calculated to 14 digits of accuracy, so if the solution were some less "convenient" number (say, $\sqrt{3}$ or $1/\pi$), we would have an answer that would be accurate enough for nearly any computation.

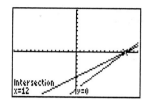

Note: An approximation for the point of intersection can be found simply by moving the TRACE cursor as near the intersection as possible. The amount of error can be minimized by "zooming in" on the graph. This is the only method available for graphing calculators such as the TI-81.

The second graphical approach is to use the *x*-**intercept method**, which seeks the *x*-coordinate of the point where a graph crosses the *x*-axis. Specifically, we want to know where the graph of y1−y2 crosses the *x*-axis, where y1 and y2 are as defined above. This is because the equation $\frac{1}{2}x - 6 = \frac{3}{4}x - 9$ can only be true when $\frac{1}{2}x - 6 - \left(\frac{3}{4}x - 9\right) = 0$. (This is the approach illustrated in the text for Example 9.)

To find this *x*-intercept, begin by defining y3=y1−y2 on the GRAPH:y(x)= screen. We could do this by re-typing the formulas entered for y1 and y2, but having typed those formulas once, it is more efficient to do this as shown on the right. The simplest way to type "y1" and "y2" is to use the [F2] key to produce "y." Note that y1 and y2 have been "de-selected" so that they will not be graphed (see section 12 of the introduction, page 38).

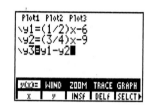

We must first select a viewing window which shows the *x*-intercept; we again use $[-15, 15] \times [-10, 10]$. The TI-86 can automatically locate this point with the GRAPH:MATH:ROOT ([GRAPH][MORE][F1][F1]) feature ("root" is a synonym for "*x*-intercept"). The TI-86 prompts for left and right bounds (numbers that are, respectively, less than and greater than the root) and a guess, then attempts to locate the root between the given bounds. (Provided there is only one root between the bounds, and the function is "well-behaved"— meaning it has some nice properties like continuity—the calculator will find it.) The screens below illustrate these steps.

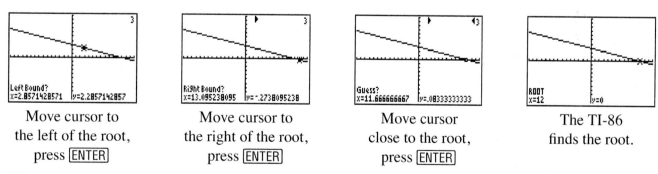

| Move cursor to the left of the root, press ENTER | Move cursor to the right of the root, press ENTER | Move cursor close to the root, press ENTER | The TI-86 finds the root. |

[85] *The TI-85* GRAPH:MATH:ROOT *feature is* [GRAPH][MORE][F1][F3]. *Unlike the TI-86, the TI-85 does not require the user to identify left and right bounds for the root.*

Section 2.6   Example 2   (page 252)   Graphing Piecewise-Defined Functions

The text shows two methods for entering piecewise-defined functions on the TI-86. Recall that the inequality symbols $>, <, \geq, \leq$ are found in the TEST menu (2nd 2). The use of DOT (DrawDot) mode is not crucial to the graphing process, as long as one remembers that vertical line segments connecting the "pieces" of the graph (in this case, at $x = 2$) are not really part of the graph.

The advantage of the method used in (b)—placing the piecewise definition in a single formula—is that y1 will act exactly like the function f. For example, to evaluate $f(5)$ in (a), one would first have to check which formula to use (y1 or y2). To evaluate $f(5)$ in (b), simply enter y1(5). Also, when a single formula is used, TRACE will properly show the behavior of the function as the trace cursor moves left and right.

Extension: Use a similar procedure for more complicated piecewise-defined functions, such as

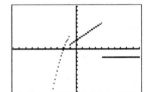

$$f(x) = \begin{cases} 4 - x^2 & \text{if } x < -1 \\ 2 + x & \text{if } -1 \leq x \leq 4 \\ -2 & \text{if } x > 4 \end{cases}$$

shown here in DOT (DrawDot) mode on the standard viewing window. It does not work to define y1=(4-x²)(x<-1)+(2+x)(-1≤x≤4)+(-2)(x>4). Instead, use

   y1=(4-x²)(x<-1)+(2+x)(-1≤x and x≤4)+(-2)(x>4),   or

   y1=(4-x²)(x<-1)+(2+x)(-1≤x)(x≤4)+(-2)(x>4)

(the "and" in the first formula is most easily entered either by selecting it from the CATALOG, or by simply typing it letter-by-letter; note the spaces before and after, typed by pressing 2nd (-)).

Section 2.6   Example 3   (page 253)   Graphing a Greatest Integer Function

We wish to graph the step function $y = \llbracket \frac{1}{2}x + 1 \rrbracket$. This is entered on the TI-86 as shown on the right (any one of y1, y2, or y3). int is found in the MATH:NUM menu, 2nd × F1.

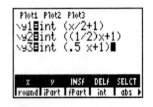

See page 39 for information on setting the thickness of a graph. Note that the symbol next to Y2 in the screen shown in the text indicates that the graph should be thick, as it does (for y2) on the TI-86.

It is possible to distinguish between the two graphs without having them drawn using different styles. When two or more graphs are drawn, the TRACE feature can be used to determine which graphs correspond to which formulas: By pressing ▲ and ▼, the trace cursor jumps from one graph to another, and the upper right corner displays the number of the current formula. On the right, the trace cursor is on graph 2—that is, the graph of $g(x) = 2|x|$.

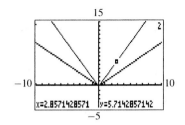

[85] *The TI-85 does not support "function notation" like* y2=y1(x+2) *and* y3=y1(x-6), *but the same thing can be accomplished by defining* y2=evalF(y1,x,x+{2,-6}).

Section 2.8   Example 5   (page 279)   Evaluating Composite Functions

The expressions shown in the calculator screens on page 279 will also work on the TI-86's home screen. To type "y" on the home screen, press 2nd ALPHA 0. Note that while commands and built-in functions like abs can be typed in in either upper- or lowercase, functions like y1 must be typed with a lowercase y.

Section 3.1   Example 1   (page 304)   Graphing Quadratic Functions

Section 3.1   Example 2   (page 305)   Graphing a Parabola by Completing the Square

Section 3.1   Example 3   (page 306)   Graphing a Parabola by Completing the Square

The TI-86 can automatically locate the vertex of the parabola (to a reasonable degree of accuracy). We illustrate this procedure for the function in Example 1(c). After defining y1=-(1/2)(x-4)²+3 and choosing a suitable window, go to the GRAPH:MATH menu (MORE F1), shown on the first screen below. Pressing MORE again brings up the menu shown in the second screen; choose F4 for an upward-opening parabola, or F5 for a downward-opening parabola (like the one in this example). The TI-86 then prompts for left and right bounds (numbers that are, respectively, less than and greater than the location of the vertex) and a guess, and the TI-86 will try to find the highest (or lowest) point on the graph between the given bounds. (For some functions, it might not always find the correct maximum or minimum, but for parabolas, it should succeed.) The screens below illustrate the procedure.

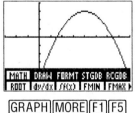

GRAPH MORE F1 F5

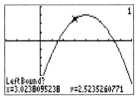

Move to the left
of the maximum,
press ENTER

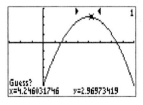

...then move the
cursor twice more.

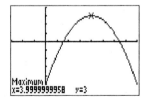

The TI-86
finds the maximum

Note that the x-coordinate rounds to 4; the degree of accuracy depends on the guess and the value of tol set on the TOLERANCE screen (2nd 3 F4). The results shown arose from tol=1E-5.

[85] *On the TI-85,* FMIN *and* FMAX *are found by pressing* GRAPH MORE F1 MORE. *The TI-85 does not prompt for left and right bounds. Also, the* TOLER *settings are accessed using* 2nd CLEAR.

Section 3.1   Example 6   (page 310)   Modeling the Number of Hospital Outpatient Visits

See page 49 for information about using the TI-86 for regression computations. On the TI-86, a quadratic regression is called a "P2Reg" (polynomial-degree-2 regression.)

Section 3.2   Example 3   (page 325)   Deciding Whether a Number is a Zero

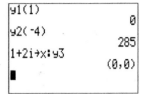

The TI-86's "function notation" evaluation method (page 48 of this manual) can also be used to check for zeros: In the screen on the right, y1, y2, and y3 have been defined (respectively) as the functions in (a), (b), and (c). While the "function notation" evaluation method fails for complex numbers—y3(1+2i) or y3(1,2) results in an error—one can evaluate y3 at $1 + 2i$ by first storing that value in x, then requesting the value of y3 (with no parentheses, which means "evaluate using the current value of x"). The screen on the right shows the output of this sequence of commands. This screen assumes that i has been defined as (0, 1), as described on page 44; the command (1,2)→x:y3 would also work.

Section 3.3   Example 6   (page 335)   Finding All Zeros of a Polynomial Function Given One Zero

Section 3.4   Example 7   (page 349)   Approximating Real Zeros of a Polynomial Function

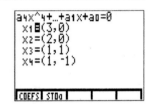

The TI-86's POLY feature (described on page 45) can determine *all* the zeros—not just the real zeros. The result of using this feature is shown on the right for Example 6, using the TI-86's complex number format.

Section 3.5   Example 2   (page 361)   Graphing a Rational Function

Note that this function is entered as y1=2/(1+x), **not** y1=2/1+x.

The issue of incorrectly drawn asymptotes is also addressed in section 12 of the introduction (page 38). Changing the window to xMin $= -5$ and xMax $= 3$ (or any choice of xMin and xMax which has -1 halfway between them) eliminates this vertical line because it forces the TI-86 to attempt to evaluate the function at $x = -1$. Since f is not defined at -1, it cannot plot a point there, and as a result, it does not attempt to connect the dots across the "break" in the graph.

Section 4.1   Example 7   (page 409)   Finding the Inverse of a Function with a Restricted Domain

Note that Figure 10 shows a square viewing window (see page 37); the mirror-image property of the inverse function would not be as clear on a non-square window. Note also that although this window is square on a TI-83, it is not quite square on a TI-86.

Section 4.2   Example 11   (page 426)   Using Data to Model Exponential Growth

See page 49 for information about using the TI-86 for regression computations.

Section 4.4   Example 1   (page 448)   Finding pH

For (a), the text shows `-log(2.5*10^(-4))`, but this could also be entered as shown on the first line of the screen on the right, since "ε" (produced with ⎣EE⎦) and "`*10^`" are nearly equivalent. The two are not completely interchangeable, however; in particular, in part (b), "`10^`" **cannot** be replaced with "ε", because "ε" is only valid when followed by an *integer*. That is, ε⁻7 produces the same result as `10^-7`, but the last line shown on the screen produces a syntax error.

(Incidentally, "`10^`" is ⎣2nd⎦⎣LOG⎦, but ⎣1⎦⎣0⎦⎣^⎦ produces the same results.)

Section 4.5   Example 8   (page 463)   Modeling Coal Consumption in the U.S.

The modeling function given in the text was found using a regression procedure (LnR, or logarithmic regression) similar to that described on page 49.

Section 5.1   Example 1   (page 495)   Solving a System by Substitution
Section 5.1   Example 3   (page 496)   Solving an Inconsistent System
Section 5.1   Example 4   (page 497)   Solving a System with Infinitely Many Solutions
Section 5.1   Example 6   (page 500)   Solving a System of Three Equations with Three Variables

The TI-86's built-in SIMULT feature can be used to solve linear systems with up to 30 unknowns. Pressing ⎣2nd⎦⎣TABLE⎦ (⎣85⎦ *or* ⎣2nd⎦⎣STAT⎦) causes the TI-86 to prompt for the number of equations, after which it prompts for the coefficients, one equation at a time. The screens below show the first and third equations being entered; the arrow keys (or ⎣F1⎦ and ⎣F2⎦) allow the user to move from one coefficient to another. When all the constants have been entered, pressing ⎣F5⎦ solves for the three unknowns (which the TI-86 calls x_1, x_2, and x_3, rather than x, y, and z).

For systems with no solutions, or infinitely many solutions (as in Example 4), the TI-86 gives a SINGULAR MAT error ("MAT" stands for "matrix"; these are discussed in Section 5.2 of the text).

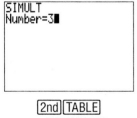

⎣2nd⎦⎣TABLE⎦
(or ⎣2nd⎦⎣STAT⎦)

Enter coefficients...   Press ⎣F5⎦ when done.

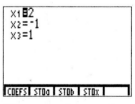

Here are the results.

Section 5.1   Example 8   (page 502)   Using Curve Fitting to Find an Equation Through Three Points

Note that the equation found in this example by algebraic methods can also be found very quickly on the TI-86 by performing a quadratic regression (P2Reg— see page 49) on the three given points. In the screen shown, PRegC is the list of "polynomial regression coefficients." These agree with the coefficients found in the text.

Section 5.2   Example 1   (page 512)   Using the Gauss-Jordan Method

Section 5.2   Example 2   (page 514)   Using the Gauss-Jordan Method

The menu shown in the text (Figure 7) is from a TI-83. The TI-86 has the four matrix row operations, but they are located in the MATRX:OPS menu (not MATRX:MATH), and the names are slightly different. The screen on the right shows this menu (visible after pressing 2nd 7 F4 MORE).

Before describing these operations, a few words about entering matrices into the TI-86: The matrix in the text has the name "[A]." On the TI-82 and TI-83, all matrices have names of the form [*letter*]; on the TI-86, matrices can have any variable name. (In fact, variable names *cannot* include the bracket characters on the TI-86.)

To change the entries in a matrix (e.g., to enter the augmented matrix for this example), either store the matrix using a home screen command (similar to the form used in the screen above), or use the MATRX:EDIT feature, then choose the name of a matrix (or type a name for a new matrix), specify the number of rows and columns, and type in the entries. The screen on the right shows the "edit" view of the matrix MAT, entered above.

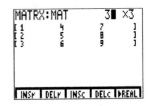

The formats for the matrix row-operation commands are:

- rSwap(*matrix*,*A*,*B*) produces a new matrix that has row *A* and row *B* swapped.

- rAdd(*matrix*,*A*,*B*) produces a new matrix with row *A* added to row *B*.

- multR(*number*,*matrix*,*A*) produces a new matrix with row *A* multiplied by *number*.

- mRAdd(*number*,*matrix*,*A*,*B*) produces a new matrix with row *A* multiplied by *number* and added to row *B* (row *A* is unchanged).

Shown below are examples of each of these operations on the matrix $\text{MAT} = \begin{bmatrix} 1 & 4 & 7 \\ 2 & 5 & 8 \\ 3 & 6 & 9 \end{bmatrix}$.

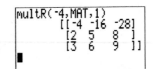
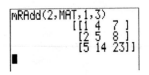

Keep in mind that these row operations leave the matrix MAT untouched. To perform a sequence of row operations, each result must either be stored in a matrix, or use the result variable Ans as the matrix. For example, with MAT equal to the augmented matrix for the system given in this example, the following screens illustrate the initial steps in the Gauss-Jordan method. Note the use of Ans in the last two screens.

```
MAT
        [[1 -1 5  -6]
         [3  3 -1 10]
         [1  3  2  5]]
■
```
```
mRAdd(-3,MAT,1,2)
        [[1 -1  5  -6]
         [0  6 -16 28]
         [1  3  2   5]]
■
```
```
mRAdd(-1,Ans,1,3)
        [[1 -1  5  -6]
         [0  6 -16 28]
         [0  4 -3  11]]
■
```
```
multR(1/6,Ans,2)▶Frac
        [[1 -1 5   -6  ]
         [0  1 -8/3 14/3]
         [0  4 -3   11 ]]
```

As we see in Figure 7(d), the rref command ([2nd][7][F4][F5]) will do all the necessary row operations at once, making these individual steps seem tedious. On the TI-86, the parentheses are not needed, so the necessary command is rref MAT.

Section 5.3   Example 1   (page 523) — Evaluating a 2×2 Determinant
Section 5.3   Example 3   (page 526) — Evaluating a 3×3 Determinant

On a TI-86, the parentheses around the matrix name are not needed. The function "det" is available in the MATRX:MATH menu ([2nd][7][F3][F1]). Shown on the right is the TI-86 version of the screen in the text.

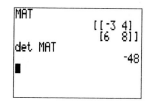

```
MAT
          [[-3 4]
           [6  8]]
det MAT
               -48
■
```

The determinant of a 3 × 3—or larger—matrix is as easy to find with a calculator as that of a 2 × 2 matrix. (At least, it is as easy for the user; the calculator is doing all the work!) Note, however, that trying to find the determinant of a non-square matrix (for example, a 3 × 4 matrix) results in a DIMENSION error.

Section 5.5   Example 2   (page 544) — Solving a Nonlinear System by Elimination

See the discussion of Example 2 from Section 2.2 (page 48 of this manual) for tips on entering formulas like these.

The text shows the intersections as found by the procedure built in to the calculator (described on page 50 of this manual, called "ISECT" on the TI-86). However, that is somewhat misleading; for these equations, the TI-86 can only find these intersections if the "guesses" supplied by the user are the exact x coordinates of the intersections (that is, -2 and 2). This is because the two circle equations are only valid for $-2 \le x \le 2$, while the hyperbola equations are only valid for $x \le -2$ and $x \ge 2$. Since only ± 2 fall in both of these domains, any guess other than these two values results in a BAD GUESS error.

Section 5.6   Example 1   (page 555) — Graphing a Linear Inequality

With the TI-86, there are two ways to shade above or below a function. The simpler way is to use the "shade above" graph style (see page 39). The screen on the right shows the "shade above" symbol next to y1, which produces the graph shown in the text.

```
Plot1 Plot2 Plot3
▶y1■-(1/4)x+1■
```

The other way to shade is the Shade(command, accessed in the GRAPH:DRAW menu ([GRAPH][MORE][F2][F1]). The format is

    Shade(*lower*,*upper*,*min X*,*max X*,*pattern*,*resolution*)

Here *lower* and *upper* are the functions between which the TI-86 will draw the shading (above *lower* and below *upper*). The last four options can be omitted. *min X* and *max X* specify the starting and ending x values for the shading. If omitted, the TI-86 uses xMin and xMax.

The last two options specify how the shading should look. *pattern* determines the direction of the shading: 1 (vertical—the default), 2 (horizontal), 3 (negative-slope 45°—that is, upper left to lower right), or 4 (positive-slope 45°—that is, lower left to upper right). *resolution* is a positive integer (1,2,3,...) which specifies how dense the shading should be (1 = shade every column of pixels, 2 = shade every other column, 3 = shade every third column, etc.). If omitted, the TI-86 shades every column; i.e., it uses *resolution* = 1.

[85]   *The TI-85 does not support graph styles, and the* Shade *command has fewer options. Specifically, the TI-85 only does "solid" shading; the* pattern *and* resolution *options are not available.*

To produce a graph like the one accompanying Example 1 in the text, the appropriate command (typed on the home screen) would be something like

    Shade(-(1/4)x+1,10,-2,6,1,2).

The use of 10 for the *upper* function simply tells the TI-86 to shade up as high as necessary; this could be replaced by any number greater than 4 (the value of yMax for the viewing window shown). Also, if the function y1 had previously been defined as -(1/4)x+1, this command could be shortened to Shade(y1,10,-2,6,1,2).

One more useful piece of information: Suppose one makes a mistake in typing the Shade(command (e.g., switching *upper* and *lower*, or using the wrong value of *resolution*), resulting in the wrong shading. The screen on the right, for example, arose from typing Shade(-(1/4)x+1,1,-2,6,1,1). In order to achieve the desired results, the mistake must be erased using the CLDRW (clear drawing) command, found in the GRAPH:DRAW menu ([GRAPH][MORE][F2][MORE]

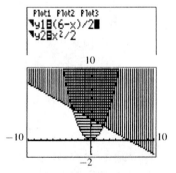

[MORE][F1]). Then—perhaps using deep recall (see page 33)—correct the mistake in the Shade(command and try again.

Section 5.6   Example 2   (page 556)   Graphing Systems of Inequalities

The easiest way to produce (essentially) the same graph as that shown in the text is to use the "shade above" graph style (see page 39). The screen on the right (above) shows the "shade above" symbol next to y1 and y2, with the results shown on the graph below. When more than one function is graphed with shading, the TI-86 rotates through the four shading patterns (see page 58); that is, it graphs the first with vertical shading, the second with horizontal, and so on. All shading is done with a resolution of 2 (every other pixel).

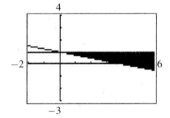

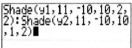

The Shade command can be used to produce this from the home screen. If y1=(6-x)/2 and y2=x²/2, the commands at right produce the graph shown in the text.

A nicer picture can be created, with a little more work, by making the observation that if y is greater than both $x^2/2$ and $(6-x)/2$, then for any x, y must be greater than the larger of these two expressions. The TI-86 provides a convenient way to find the larger of two numbers with the max(function, located in the LIST:OPS menu ([2nd][–][F5][F5]). Then max(y1,y2) will return the larger of y1 and y2, and the "y equals" screen entries shown on the right (above) will produce the graph shown below it. (Note the graph style settings: y1 and y2 display as solid curves, while y3 has the "shade above" style.) The home-screen command Shade(max(y1,y2),11,-10,10,1,2) would produce similar results.

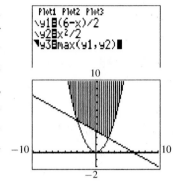

Extension: The table below shows how (using graph styles or home-screen commands) to shade regions that arise from variations on the inequalities in this example, assuming that y1=(6-x)/2 and y2=x²/2. (The results of these commands are not shown here.)

For the system...	or equivalently...	the command would be...
$x < 6 - 2y$ $x^2 < 2y$	$y < (6-x)/2$ $y > x^2/2$	shade below y1 and above y2, or enter Shade(y2,y1,-10,10,1,2)
$x > 6 - 2y$ $x^2 > 2y$	$y > (6-x)/2$ $y < x^2/2$	shade above y1 and below y2, or enter Shade(y1,y2,-10,10,1,2)
$x < 6 - 2y$ $x^2 > 2y$	$y < (6-x)/2$ $y < x^2/2$	shade below y1 and below y2, or enter Shade(-10,min(y1,y2),-10,10,1,2)

[85] *Because the TI-85 does not have graph styles, one must use the* Shade *commands given above, combined with* min *or* max, *to produce the appropriate graphs.*

Section 5.7   Example 2   (page 566)   Adding Matrices

Section 5.7   Example 3   (page 567)   Subtracting Matrices

Section 5.7   Example 4   (page 568)   Multiplying Matrices by Scalars

Section 5.7   Example 6   (page 571)   Multiplying Matrices

See page 56 for information about matrices on the TI-86.

Section 5.8   Example 1   (page 580)   Verifying the Identity Property of I_3

On the TI-86, the function to produce identity matrices is ident, which does not require parentheses, and is located in the MATRX:OPS menu ([2nd][7][F4][F3]). The screen on the right shows the 3×3 identity matrix.

| Section 5.8 | Example 2 | (page 583) | Finding the Inverse of a 3×3 Matrix |
| Section 5.8 | Example 3 | (page 584) | Identifying a Matrix with No Inverse |

An inverse matrix is found using 2nd EE —the same method used to find the inverse of a real number. For example, to find the inverse of matrix MAT, type MAT (or choose that name from the MATRX:NAMES menu) then press 2nd EE ENTER . An error will occur if the matrix is not square, or if it is a singular matrix (as in Example 3).

Section 6.1	Example 1	(page 607)	Graphing a Parabola with Horizontal Axis
Section 6.1	Example 2	(page 608)	Graphing a Parabola with Horizontal Axis
Section 6.2	Example 1	(page 618)	Graphing Ellipses Centered at the Origin
Section 6.3	Example 1	(page 629)	Using Asymptotes to Graph a Hyperbola

See the discussion of Example 2 from Section 2.2 (page 48 of this manual) for tips on entering formulas like these.

When entering these formulas, make sure to use enough parentheses, so that operations are performed in the correct order; e.g., for Example 2 of Section 6.1, the formula in y1 should look like this:

```
y1=-1.5+√((x-.5)/2)
```

Also, when looking at a calculator graph like the one shown in Figure 18 (page 619), keep in mind that the apparent gap between this graph and the x-axis is not really there; it is a flaw that arises from the calculator's method of plotting functions (see discussion on page 39 of this manual).

| Section 7.1 | Example 3 | (page 655) | Modeling Insect Population Growth |

One way to produce this sequence (and similar recursively defined sequences) is with the Ans variable, as shown on the right. (After pressing 1 ENTER , then typing the next line and pressing ENTER , we simply press ENTER over and over to generate successive terms in the sequence.)

Figure 4 of the text shows a table of sequence terms, while Figure 5(b) shows a plot of the terms. While the TI-86 can generate tables for functions, it does not have the TI-83's "sequence mode," which was used to create these screens. See the discussion of Example 7 from Section 7.3 (below) for information on creating a plot like Figure 5(b), but note that on the TI-86, recursive sequences are tedious to plot. (Non-recursive sequences are a bit easier.)

Section 7.1   Example 4   (page 657)   Using Summation Notation
Section 7.2   Example 9   (page 669)   Using Summation Notation

The `seq(` command can be found in the LIST:OPS menu ([2nd][−][F5][MORE][F3]).
Given a formula a_n for the nth term in a sequence, the command

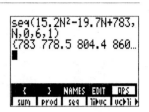

    `seq(`*formula*`,`*variable*`,`*start*`,`*end*`,`*step*`)`

produces the list $\{a_{start}, a_{start+step}, \ldots, a_{end}\}$. If *step* is omitted, a value of 1 is
assumed. The size of the resulting list can have no more than 2800 items (or
fewer, as available memory allows).

Note that *variable* can be any letter (or letters)—K and I are used in the text, but
x is more convenient (since it can be typed with [x-VAR]).

The `sum` command can be found in the LIST:OPS menu ([2nd][−][F5][MORE][F1]). On
the TI-86, the parentheses are not required. `sum` can be applied to any list—either
to a list variable (called "`list`" in the example screen on the right), or directly to
a list created with the `seq(` command.

Note: On the TI-86, list variables can be given any kind of legal variable name (see page 34).

Section 7.3   Example 7   (page 676)   Summing the Terms of an Infinite Geometric Series

To produce the plot from Figure 10 on the TI-86, begin by storing values in the lists
xStat and yStat as shown on the right, so that xStat contains values of n, and
yStat contains the corresponding values of S_n. (The slightly shorter command
`(1-(1/3)^xStat)/(2/3)→yStat` has the same effect as the second command
shown.) This step is tedious for recursively defined sequences, but fairly simple
in this case.

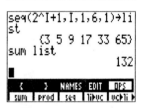

Next press [2nd][+][F3] to bring up the STAT:PLOT menu, shown on the right. Select
`Plot1` by pressing [F1] (or choose one of the other two plots).

Make the settings shown on the screen on the right. For `Type`, choose [F1] (SCAT)
for a scatterplot.

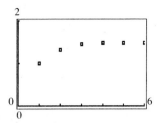

Set up the viewing window (GRAPH:WINDOW) to match that shown in the text. Also, check that nothing else will be plotted: Press GRAPH F1 and make sure that all the equals signs are not highlighted, and that only Plot1 is highlighted. If Plot2 or Plot3 is on, use the arrow keys to move the cursor up to that plot, then press F5 (SELCT). Finally, press GRAPH F5 to produce the graph.

Note: When finished with a statistics plot like this one, it is a good idea to turn it off so that the TI-86 will not attempt to display it the next time GRAPH is pushed. This can be done using the SELCT option on the GRAPH:y(x)= screen, or by executing the PlOff command, by pressing 2nd + F3 F5 ENTER .

Section 7.4   Example 1   (page 687)   Evaluating Binomial Coefficients

Section 7.6   Example 4   (page 701)   Using the Permutations Formula

---

The nCr and nPr commands are found in the MATH:PROB menu (2nd × F2). (Also located in that menu is the factorial operator "!.")

Section 7.7   Example 6   (page 716)   Finding Probabilities in a Binomial Experiment

---

The TI-86 does not have statistical distribution functions like binompdf. The computations shown in this example must be done by manually entering the entire binomial probability formula (or by obtaining a program to automate such computations.)

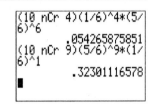

Introduction

The information in this section is essentially a summary of material that can be found in the TI-89 manual. Consult that manual for more details.

Owners of a TI-92 will find that most of this material applies to that calculator as well.

1   Power

To power up the calculator, simply press the $\boxed{\text{ON}}$ key. The screen displayed at this point depends on how the TI-89 was last used. It may show the "home screen"—a menu (the toolbar) across the top, the results of previous computations (if any) in the middle (the history area), a line for new entries (which may be blank, or may show the previous entry), and a status line at the bottom. An example of this home screen is shown on the right.

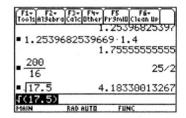

The TI-89 may show some other screen—perhaps a graph, an error message, a menu, or something else. If so, one can return to the home screen by pressing either the $\boxed{\text{HOME}}$ key (from a graph), or the $\boxed{\text{ESC}}$ key (from an error message), or $\boxed{\text{2nd}}\boxed{\text{ESC}}$ (from a menu). Note that the "second function" of $\boxed{\text{ESC}}$—written in yellow type above the key—is "QUIT."

If the screen is blank, or is too dark to read, one may need to adjust the contrast (see the next section).

To turn the calculator off, press $\boxed{\text{2nd}}\boxed{\text{ON}}$ (OFF), in which case the TI-89 will start up at the home screen next time. This will not work if an error message is displayed. Pressing $\boxed{\bullet}\boxed{\text{ON}}$ will also turn the TI-89 off, but when $\boxed{\text{ON}}$ is next pressed, the screen will show exactly what it showed before. The calculator will automatically shut off if no keys are pressed for several minutes, in which case it will behave as if $\boxed{\bullet}\boxed{\text{ON}}$ had been pressed.

2   Adjusting screen contrast

If the screen is too dark (all black), decrease the contrast by pressing and holding $\boxed{\bullet}$ and $\boxed{-}$. If the screen is too light, increase the contrast by pressing and holding $\boxed{\bullet}$ and $\boxed{+}$. If the screen never becomes dark enough to see, the batteries should be replaced.

3   Replacing batteries

To replace the four AAA batteries, first turn the calculator off (2nd ON), then remove the back cover, remove and replace each battery, replace the back cover, then turn the calculator on again. (After replacing batteries, one may need to adjust the contrast down as described above.) Note: The status line at the bottom of the home screen should display BATT when the batteries are getting low.

4   Basic operations

Simple computations are entered in essentially the same way they would be written. For example, to compute $2 + 17 \times 5$, press 2+17×5 ENTER (the ENTER key tells the calculator to act on what has been typed). Standard order of operations (including parentheses) is followed. Note that the entry and the result (87) are displayed in the last line of the history area, and the entry is also displayed (and highlighted) in the entry line. If any new text is typed, this highlighted text will be deleted.

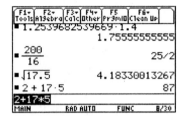

The result of the most recently entered expression is stored in ans(1), which is typed by pressing 2nd (-) (the word "ANS" appears in yellow above this key). For example, 5+2nd(-) ENTER will add 5 to the result of the previous computation. Note that in the history area, "ans(1)" has been replaced by "87."

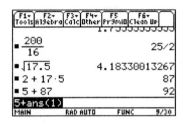

After pressing ENTER, the TI-89 automatically produces ans(1) if the first key pressed is one which requires a number before it; the most common of these are +, −, ×, ÷, ^, and STO▸. For example, +5 ENTER would accomplish the same thing as the keystrokes above (that is, it adds 5 to the previous result). Again, note that the history area shows the value of ans(1) rather than the text "ans(1)."

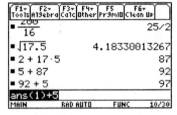

Although the previously entered expression disappears from the entry line if anything is typed, that expression can be re-evaluated by simply pressing ENTER. This can be especially useful in conjunction with ans(1).

Several expressions can be evaluated together by separating them with colons (2nd 4). When ENTER is pressed, the result of the *last* computation is displayed. (The other results are lost. An example showing how this can be used is shown later.)

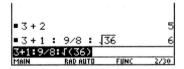

5   Cursors

When typing, the appearance of the cursor and the status line indicates the behavior of the next keypress.

The cursor appears as either a flashing vertical line (the default) or a flashing solid block. The vertical line indicates that subsequent keypresses will be *inserted* at the current cursor location. The block cursor indicates that subsequent keypresses will *overwrite* the character(s) to the right of the cursor. (Of course, if the cursor is located at the right end of the entry line text, these two behaviors are equivalent.) To switch between between these two modes of operation, press [2nd][←] (INS).

By default, most keys produce the character shown on the key itself. The four modifier keys [2nd], [♦], [↑], and [alpha] change this. Pressing any one of these keys causes a corresponding indicator to appear in the status line, and the next keypress will then do something different from its primary function. Pressing [2nd] or [♦] causes the next keypress to produce the results—the character or operation—indicated by (respectively) the yellow or green text above that key. If [2nd] or [♦] is pressed by mistake, pressing it a second time will undo that modifier.

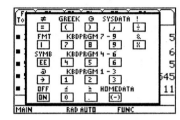

For [2nd], that makes each key's function fairly clear, but many of the keys have no green text above them, leading one to think that the [♦] modifier would accomplish nothing with that key. In fact, nearly every key does something in response to the [♦] key. For an easy way to see the [♦] functions of the lower half of the keypad, press [♦][EE], which produces the display on the right. (Note that the letter associated with [EE] is "K"—think "K" for "keys.") For example, [♦] followed by [=], [)], [÷], [×], [STO▸], [0], or [.] produces the character shown. [♦][(] followed by any letter produces the Greek equivalent of that letter (or as near an equivalent as there is); e.g., [♦][(][Z] produces ζ ("zeta"). [♦][[] allows one to change the number of previous entries saved in the history area. The other [♦] functions are beyond the scope of this manual.

Pressing [alpha] means that the next keypress will produce the (lowercase) letter or other character printed in purple above that key (if any). [alpha] has no effect on (and is not needed for) [X], [Y], [Z], and [T]. Following that letter, subsequent keypresses will produce their primary functions (i.e., not letters). To produce an uppercase letter, press [alpha][↑] followed by a letter key. ([alpha] is not needed for [X], [Y], [Z], and [T].)

The TI-89 can be "locked" into (lowercase) alphabetic mode by pressing [2nd][alpha] (or [alpha][alpha]). From then on, each key produces its letter. This continues until [alpha] is pressed again, which takes the TI-89 out of alphabetic mode.

To lock the TI-89 in uppercase alphabetic mode, press [↑][alpha]. As before, pressing [alpha] again takes the TI-89 out of alphabetic mode.

6   Accessing previous entries ("deep recall")

By repeatedly pressing [2nd][ENTER] (ENTRY), previously typed expressions can be retrieved for editing and re-evaluation. Pressing [2nd][ENTER] once recalls the most recent entry; pressing [2nd][ENTER] again brings up the second most recent, etc.

More conveniently, pressing ⊝ and ⊝ allows one to select previous entries and results from the history area; simply highlight the desired expression and press [ENTER]. For example, pressing ⊝⊝[ENTER] would copy the previous entry to the new-entry line, while ⊝⊝⊝[ENTER] would copy the second-previous *result*.

Once an expression is on the new-entry line at the bottom of the home screen, it can be edited in various ways. Text can be deleted (using [←] to delete the character before the cursor, or [♦][←] to delete the character after the cursor). New text can be inserted (see the previous section). One can even highlight text by pressing and holding [↑] together with an arrow key, then cut ([♦][2nd]) or copy ([♦][↑]) it to a "clipboard," so that it can be pasted ([♦][ESC]) somewhere else in the expression.

7   Menus

Certain keys and key combinations bring up a menu in a window with a variety of options. Shown is the menu produced by pressing F2 (Algebra) from the home screen. The arrow next to "8" means that there are more options available (which can be seen by pressing ⊙ or ⊙). To select one of these options (and paste the corresponding command on the entry line), simply press the number (or letter) next to the option. Alternatively, use ⊙ and ⊙ to highlight the desired option and press ENTER.

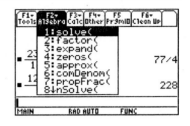

2nd 5 brings up the MATH menu shown on the right (above). For each of these menu options, the triangle ("▸") on the right side indicates that selecting that option brings up a sub-menu. Below on the right is the List sub-menu (option 3 of the MATH menu). Note that the status line

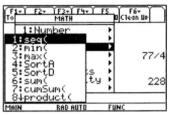

contains a reminder of how to use these menus; ESC can be used to exit from one level of a menu (and 2nd ESC would remove all menus and return to the home screen).

The various commands in these menus are too numerous to be listed here. They will be mentioned as needed in the examples.

One last comment is worthwhile, however. Some functions that may be used frequently are buried several levels deep in the menus, and may take many keystrokes to access. Worse, the location of the function might be forgotten (is it in the Algebra or MATH:Number menu?), necessitating a search through the menus. It is useful to remember three things:

- Any command can be typed one letter at a time, in either upper- or lowercase; e.g., ↑ ALPHA = ((3 ALPHA (will type the letters "ABS(", which has the same effect as 2nd 5 1 2.

- Any command can be found in the CATALOG menu. Since the commands appear in alphabetical order, it may take some time to locate the desired function. Pressing any letter key (it is not necessary to press alpha first) brings up commands starting with that letter; e.g., pressing 9 brings up the list on the right, while pressing − shows commands starting with "O."

- An alternate home-screen menu bar can be found by pressing 2nd HOME (CUSTOM). Pressing 2nd HOME again toggles back to the "standard" home-screen menu bar. It is possible to change the commands listed in this CUSTOM menu, but the process is somewhat tedious. Full details can be found in the TI-89 manual, but as an example: The command

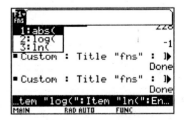

        Custom:Title "Fns":Item "abs(":Item "log(":Item "ln(":EndCustm

would result in a CUSTOM menu bar for which F1 would produce the menu shown on the right.

8   Variables

The letters A through Z (upper- or lowercase), and also sequences of letters (like "High" or "count") can be used as variables (or "memory") to store numerical values. To store a value, type the number (or an expression) followed by STO▸, then a letter or letters (pressing alpha if necessary), then ENTER. That variable name can then be used in the same way as a number, as demonstrated at right. These variable names are *not* case-sensitive, so "A" is the same as "a," and "CoUnT" is the same as "count."

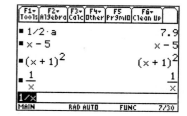

Note: The TI-89 interprets 2a as "2 times a"—the "*" symbol is not required (this is consistent with how we interpret mathematical notation). As for order of operations, this kind of multiplication is treated the same as "*" multiplication (see the screen above).

If a variable is used for which no value has been assigned, it is treated as an unknown value, and expressions involving it remain unevaluated. In the screen on the right, the variable x has no assigned value. Any variable's assigned value can be erased ("forgotten") by issuing the command DelVar (F4 4 from the home screen menubar) followed by the variable name. All one-letter variables can be cleared by choosing 2nd F1 (Clean Up) 1, or using the NewProb command (2nd F1 2), which also clears the history area.

9   Setting the modes

By pressing MODE, one can change many aspects of how the calculator behaves. For most of the examples in this manual, the MODE settings will be as shown on the three screens below (although in some cases the settings are not crucial). Some (not all) of these options are described below; consult the TI-89 manual for more details. Changes in the settings are made using the arrows keys and ENTER.

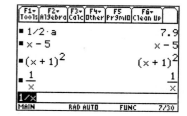

The Graph setting can be either FUNCTION, PARAMETRIC, POLAR, SEQUENCE, 3D, or DIFF EQUATIONS. This setting determines whether formulas to be graphed are functions (*y* as a function of *x*), parametric equations (*x* and *y* as functions of *t*), polar equations (*r* as a function of *θ*), or sequences (*u* as a function of *n*); the other two settings are beyond the scope of this manual. The text accompanying this manual uses FUNCTION and SEQUENCE modes. The current value of this setting is indicated in the status line at the bottom of the calculator screen—FUNC, PAR, etc.

The Display Digits setting can be FLOAT, FIX *n*, or FLOAT *n*, where *n* is an integer from 1 to 12. This specifies how numbers should be displayed: FLOAT means that the TI-89 should display all non-zero digits (up to a maximum of 12), while (e.g.) FLOAT 4 means that a total of 4 digits will be displayed. Meanwhile, FIX 4 means that the TI-89 will attempt to display 4 digits beyond the decimal point.

Angle can be either RADIAN or DEGREE, indicating whether angle measurements should be assumed to be in radians or degrees. (A right angle measures $\frac{\pi}{2}$ radians, which is equivalent to $90°$.) The current value of this setting is indicated in the status line at the bottom of the calculator screen—RAD or DEG. This text does not refer to angle measurement.

The Exponential Format setting is either NORMAL, SCIENTIFIC, or ENGINEERING; this specifies how numbers should be displayed. The screen on the right shows the number "12345." displayed in each of these modes: NORMAL mode displays numbers in the range $\pm999,999,999,999$ with no exponents, SCIENTIFIC mode displays all numbers in scientific notation, and ENGINEERING mode uses only exponents that are multiples of 3. Note: "E" is short for "times 10 to the power," so $1.2345E4 = 1.2345 \times 10^4 = 1.2345 \times 10000 = 12345$.

The Complex Format is either REAL, RECTANGULAR, or POLAR, and specifies the display mode for complex numbers. REAL means that the TI-89 will produce an error if an expression requires the computation of (e.g.) a square root of a negative number, and the other two settings determine whether complex results should be displayed in rectangular or polar format. The text uses only the first of these formats. More information about complex numbers can be found beginning on page 78 (Section 1.3, Example 1) of this manual.

The Vector Format setting (RECTANGULAR, CYLINDRICAL, or SPHERICAL) indicates the default display format for vectors (not used in this text).

Pretty Print (ON or OFF) determines how expressions (input and output) should be displayed. The first two entries on the right were performed with Pretty Print on, and the other two were done with Pretty Print off.

The Exact/Approx setting (AUTO, EXACT, or APPROXIMATE) determines whether results should be considered to be exact or approximate. EXACT means that all decimals are converted to fractions; e.g., .9 is displayed as $9/10$, and $\int(2.5)$ is $\sqrt{10}/2$. APPROXIMATE means that everything is converted to decimal form; for example, $\int(2.5)$ produces $1.5811\ldots$. With the AUTO setting, the TI-89 decides whether to display a result as exact or approximate based on whether there is a decimal point in the entry—for example, $\int(5/2)$ yields $\sqrt{10}/2$, while $\int(2.5)$ yields $1.5811\ldots$. **Note:** Regardless of this setting, pressing ◆ENTER instead of ENTER to process an entry will cause the TI-89 to show a decimal (approximate) result. (The current value of this setting is indicated in the status line at the bottom of the calculator screen—AUTO, EXACT, or APPROX.)

The other mode settings deal with issues that are beyond the scope of the textbook, and are not discussed here.

10   Setting the graph window

Pressing ◆F2 brings up the WINDOW settings. The exact contents of the WINDOW menu vary depending on the Graph mode setting; below are six examples showing this menu in each possible Graph modes. (The last two are not used in this manual, but are shown here for reference.)

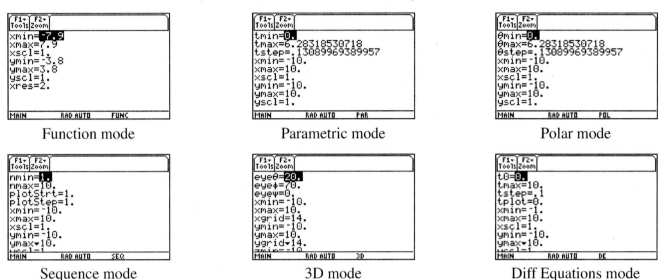

Function mode   Parametric mode   Polar mode

Sequence mode   3D mode   Diff Equations mode

All these menus include the values xmin, xmax, xscl, ymin, ymax, and yscl. When ◆F3 (GRAPH) is pressed, the TI-89 will show a portion of the Cartesian (*x*-*y*) plane determined by these values. In function mode, this menu also includes xres, the behavior of which is described in section 12 of this manual (page 71). The other settings in this screen allow specification of the smallest, largest, and step values of *t* (for parametric mode) or *θ* (for polar mode), or initial conditions for sequence mode.

With settings as in the "Function mode" screen shown above, the TI-89 would display the screen at right: *x* values from −7.9 to 7.9 (that is, from xmin to xmax), and *y* values between −3.8 to 3.8 (ymin to ymax). Since xscl = yscl = 1, the TI-89 places tick marks on both axes every 1 unit; thus the *x*-axis ticks are at −7, −6, . . . , 6, and 7, and the *y*-axis ticks fall on the integers from −3 to 3. This window is called the "decimal" window, and is most quickly set by pressing ◆F2F2 4 (Zoom:ZoomDec).

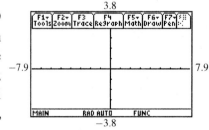

Below are three more sets of window settings, and the graph screens they produce. Note that the first graph on the left has tick marks every 10 units on both axes. The second window is called the "standard" viewing window, and is most quickly set by pressing ◆F2F2 6 (Zoom:ZoomStd). The setting $yscl = 0$ in the final graph means that no tick marks are placed on the y-axis.

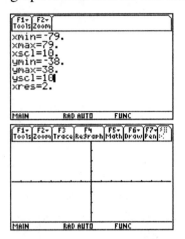 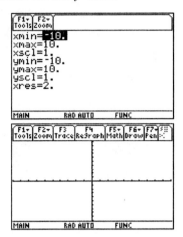 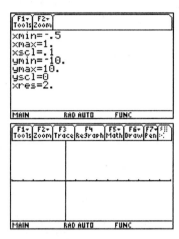

11   The graph screen

The TI-89's graph screen—that is, the portion of the screen used to display graphs, below the menu bar and above the status line—is made up of an array of rectangular dots (pixels) with 77 rows and 159 columns. All the pixels in the leftmost column have x-coordinate xmin, while those in the rightmost column have x-coordinate xmax. The x-coordinate changes steadily across the screen from left to right, which means that the coordinate for the nth column (counting the leftmost column as column 0) must be $xmin + n\Delta x$, where $\Delta x = (xmax - xmin)/158$. Similarly, the nth row of the screen (counting up from the bottom row, which is row 0) has y-coordinate $ymin + n\Delta y$, where $\Delta y = (ymax - ymin)/76$.

It is not necessary to memorize the formulas for Δx and Δy. Should they be needed, they can be determined by pressing ◆F3 (GRAPH) and then the arrow keys. When pressing ⊙ or ⊙ successively, the displayed x-coordinate changes by Δx; meanwhile, when pressing ⊙ or ⊙, the y-coordinate changes by Δy. Alternatively, the values can be found by typing "Δx" and "Δy" on the home screen; this is most easily done by pressing 2nd + 1 5 to access the CHAR:Greek menu and type the "Δ" character, then press X or Y. This produces results like those shown on the right; the values of Δx and Δy there are those for the standard viewing window.

In the decimal window $xmin = -7.9$, $xmax = 7.9$, $ymin = -3.8$, $ymax = 3.8$, note that $\Delta x = 0.1$ and $\Delta y = 0.1$. Thus, the individual pixels on the screen represent x-coordinates $-7.9, -7.8, -7.7, \ldots, 7.7$, $7.8, 7.9$ and y-coordinates $-3.8, -3.7, -3.6, \ldots, 3.6, 3.7, 3.8$. This is where the decimal window gets its name.

Windows for which $\Delta x = \Delta y$, such as the decimal window, are called square windows. Since there are just over twice as many columns as rows on the graph screen, this means that square windows should have $xmax - xmin$ just over twice as big as $ymax - ymin$. Any window can be made square be pressing ◆F2F2 5 (Zoom:ZoomSqr). To see the effect of a square window, observe the two pairs of graphs below. In each pair, the first graph is on the standard window, and the second is on a square window (after pressing ◆F2F2 5). (This changes xmin and xmax to about -20.8 and 20.8, respectively, while ymin and ymax remain unchanged at -10 and 10.) The first pair shows the line $y = x$; on the square window, this line (correctly) appears to make a $45°$ angle with the x- and y-axes. The second pair shows the lines $y = 2x - 3$

and $y = 3 - \frac{1}{2}x$; note that on the square window, these lines look perpendicular (as they should). Finally, the last pair shows a circle centered at the origin with a radius of 8. On the standard window, this looks like an oval since the screen is wider than it is tall. (The reason for the gaps in the circle will be addressed in the next section.)

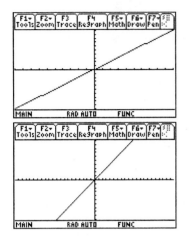 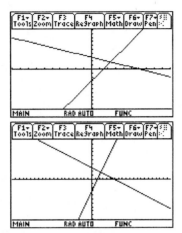 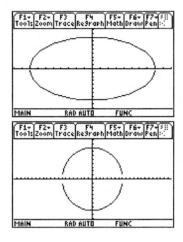

12   Graphing a function

This introductory section only addresses creating graphs in function mode. This textbook does not include examples of parametric and polar graphs; however, procedures for creating such graphs are very similar.

To see the graph of $y = 2x - 3$, begin by entering the formula into the calculator. This is done by pressing ⬥[F1] to access the "y equals" screen of the calculator. Enter the formula as y1 (or any other yn). (If y1 already has a formula, press [ENTER], [F3], or [CLEAR] first, then type the new formula.) If another y variable has a formula, position the cursor on that line and press either [CLEAR] (to delete the formula) or [F4]. The latter has the effect of

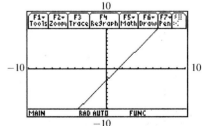

toggling the check mark for that line; which tells the TI-89 whether or not to graph that formula. In the screen on the right, only y1 will be graphed.

The next step is to choose a viewing window. See the previous section for more details on this. This example uses the standard window ([F2][6]).

If the graph has not been displayed, press ⬥[F3], and the line should be drawn. In order to produce this graph, the TI-89 considers 159 values of x, ranging from xmin to xmax in steps of Δx (assuming that xres = 1; see below for other possibilities). For each value of x, it computes the corresponding value of y, then plots that point (x, y) and draws a line between this point and the previous one. (See also the information about graph styles later in this section.)

If xres is set to 2, the TI-89 will only compute y for every other x value; that is, it uses a step size of 2Δx. Similarly, if xres is 3, the step size will be 3Δx, and so on. Setting xres higher causes graphs to appear faster (since fewer points are plotted), but for some functions, the graph may look "choppy" if xres is too large, since detail is sacrificed for speed.

Note: If the line does not appear, or the TI-89 reports an error, double-check all the previous steps. Also, check the mode settings (discussed in section 9, page 67).

Once the graph is visible, the window can be changed using ⬦ F2 (WINDOW) or F2 (ZOOM). Pressing F3 (TRACE) brings up the "trace cursor," and displays the x- and y-coordinates for various points on the line as the ⬅ and ➡ keys are pressed. (These variables—xc and yc—can also be referenced from the home screen; that is, typing xc ENTER on the home screen would show the value 2.40506....) Tracing beyond the left or right columns causes the TI-89 to adjust the values of xmin and xmax and redraw the graph.

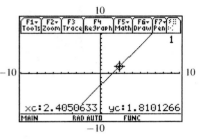

To graph the function

$$y = \frac{1}{x-3},$$

enter that formula into the "y equals" screen (note the use of parentheses on the entry line). As before, this example uses the standard viewing window.

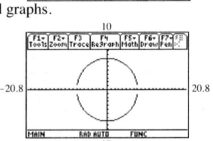

For this function, the TI-89 produces the graph shown on the right. This illustrates one of the pitfalls of the connect-the-dots method used by the calculator: The nearly-vertical line segment drawn at $x = 3$ *should not be there*, but it is drawn because the calculator connects the points

$$x \approx 2.911, \ y \approx -11.286 \ \text{ and } \ x \approx 3.165, \ y \approx 6.077.$$

Calculator users must learn to recognize these flaws in calculator-produced graphs.

The graph of a circle centered at the origin with radius 8 (shown on a square window, with xres = 1) shows another problem that arises from connecting the dots. When $x = -8.157895$, y is undefined, so no point is plotted (that is, there is no point on this circle that has x-coordinate less than -8, or greater than 8). The next point plotted on the upper half of the circle is $x = -7.894737$ and $y = 1.2934953$; since no point had been plotted for the previous x-coordinate, this is not connected to anything, so there appears to be a gap between the circle and the x-axis. The calculator is not "smart" enough to know that the graph should extend from -8 to 8.

One additional feature of graphing with the TI-89 is that each function can have a "style" assigned to its graph. To see this style, go to the "y equals" screen and press 2nd F1 (Style); the check mark indicates which style applies to the current function. These options are shown on the right; "Line," the default, means that the calculator should draw lines between the plotted points. More information can be found in the examples, and complete details are in the TI-89 manual.

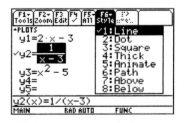

13   Adding programs to the TI-89

The TI-89's capabilities can be extended by downloading or entering programs into the calculator's memory. Instructions for writing a program are beyond the scope of this manual, but programs written by others and downloaded from the Internet (or obtained as printouts) can be transferred to the calculator in one of three ways:

1. If one TI-89 already has a program, it can be transferred to another using the calculator-to-calculator link cable. To do this, first make sure the cable is firmly inserted in both calculators. On both calculators, press [2nd][−] (VAR-LINK). On the sending calculator, use the arrow keys and [F4] to place check marks next to the programs to be transferred. On the receiving calculator, press [F3][2] (Link:Receive), then on the sending calculator, press [F3][1] (Link:Send to TI-89/TI-92 Plus).

2. If a computer with the TI-Graph Link is available, and the program file is on that computer (e.g., after having been downloaded from the Internet), the program can be transferred to the calculator using the TI Connect (or TI Graph Link) software. This transfer is done in a manner similar to the calculator-to-calculator transfer described above; specific instructions can be found in the documentation that accompanies the software. (They are not given here because of slight differences between platforms and software versions.)

3. View a listing of the program and type it in manually. (**Note:** Even if the TI-Graph Link cable is not available, the software can be used to view program listings on a computer.) While this is the most tedious method, studying programs written by others can be a good way to learn programming. To enter a program, start by choosing [APPS][7] [3] (Program Editor:New), then specify whether this is a program or a  function, and give it a name (up to eight characters, like "quadform" or "midpoint")—note that the TI-89 is automatically put into alpha mode while typing this name. Then press [ENTER] (OK), enter the commands in the program or function, and press [2nd][ESC] (QUIT) to return to the home screen when finished.

To run the program, make sure there is nothing on the current line of the home screen, then type the name of the program or function (this name is not case sensitive). Follow this with a set of parentheses, containing any required arguments, then press [ENTER]. If the program was entered manually (option 3 above), errors may be reported; in that case, press [ENTER] (GOTO), correct the mistake and try again.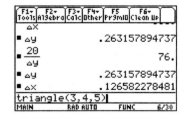

Programs can be found at many places on the Internet, including:

- `http://www.bluffton.edu/~nesterd`—the Web site of the author of this manual;

- `http://tifaq.calc.org`—a "Frequently Asked Questions" page maintained by Ray Kremer; and

- `http://www.ticalc.org`.

Additionally, one can install a variety of "APPs"—applications (programs) which can extend the capabilities of the calculator in various ways. APPs can be viewed by pressing APPS 1, or ◆ APPS; the TI-89 Titanium edition has a number of APPs preloaded, while the standard TI-89 has none. Shown is a TI-89 with four APPs installed, some of which are discussed in this manual. These and other APPs can be downloaded from education.ti.com, then installed using a Graph Link cable.

Examples

Here are the details for using the TI-89 for several of the examples from the textbook. Also given are the keystrokes necessary to produce some of the commands shown in the text's examples. In some cases, some suggestions are made for using the calculator more efficiently.

Throughout this section, it is assumed that the textbook is available for reference. The problems from the text are not restated here, and there are frequent references to the calculator screens shown in the text.

Section R.2   Example 2   (page 10)   Evaluating Exponential Expressions

---

This example discusses how to evaluate 4^3, $(-6)^2$, -6^2, $4 \cdot 3^2$, and $(4 \cdot 3)^2$.

To evaluate 4^3 with a TI-89, type 4 ^ 3 ENTER. While all exponents are entered with the ^ key, note that they are displayed in the history area as superscripts. All other computations in this example are straightforward with the TI-89, except for this *caveat*: For $(-6)^2$ and -6^2, one must use the (-) key rather than the - key to enter "negative 6." In the screen shown, the second entry in the history area was typed as ((-) 6) ^ 2, while the entry at the bottom was typed as (- 6) ^ 2. When ENTER is pressed, this entry produces a syntax error—the calculator's way of saying that the line makes no sense.

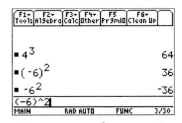

Section R.2   Example 3   (page 11)   Using Order of Operations

---

Using the calculator for parts (a) and (b) of this example is straightforward; the expressions are entered as printed. For (c) and (d), we have the expressions

$$\frac{4+3^2}{6-5 \cdot 3} \quad \text{and} \quad \frac{-(-3)^3 + (-5)}{2(-8) - 5(3)}.$$

Entering these on the TI-89 requires us to add some additional parentheses to make sure the correct order of operations is followed.

The history area on the right shows the results for (c) and (d). The entry for (c) looks the same as the printed appearance in the text, but note that it was entered as (4+3^2)/(6-5*3). For (d), note the expression was entered nearly the same as it was printed, except for the addition of parentheses around the numerator and denominator (see the entry line of the screen), but in the history area, the TI-89 has dropped the unnecessary parentheses

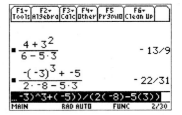

around -5 in the numerator, while in the denominator, the parentheses were exchanged for the multiplication symbol.

The entry line on the right shows a quicker way of entering the expression for (d), based on the TI-89's modification of the original entry. The extra parentheses in the original make it somewhat harder to read (and also mean more opportunities to make a mistake).

| Section R.2 | Example 7 | (page 15) | Evaluating Absolute Value |
| Section R.2 | Example 9 | (page 16) | Evaluating Absolute Value Expressions |

The abs(function of the TI-89 is in the MATH:Number menu (2nd 5 1 2). Note that while this function is entered as abs(, it is displayed in the history area using the "bar" notation.

Section R.3	Example 1	(page 22)	Using the Product Rule
Section R.3	Example 2	(page 22)	Using the Power Rules
Section R.3	Example 4	(page 25)	Adding and Subtracting Polynomials
Section R.3	Example 5	(page 26)	Multiplying Polynomials
Section R.3	Example 9	(page 29)	Dividing Polynomials

Provided the variables used $(x, y, z, \text{etc.})$ have no values assigned to them, the TI-89 can simplify expressions like these. If any values have been assigned, use the Clear a-z command (2nd F1 1), or DelVar (in the CATALOG) to remove those values.

Shown are sample entries for these examples. To multiply polynomials, use the expand command, found in the Algebra menu.

| Section R.4 | Example 1 | (page 34) | Factoring Out the Greatest Common Factor |
| Section R.4 | Example 2 | (page 35) | Factoring by Grouping |

The factor command, from the Algebra menu, can handle many problems like these.

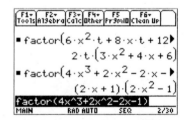

| Section R.6 | Example 4 | (page 55) | Using the Definition of $a^{1/n}$ |

In evaluating these fractional exponents with a calculator, be sure to put parentheses around the fractions. Note for (d), where the text notes that $(-1296)^{1/4}$ is not a real number, the TI-89 gives a complex result if the complex format setting (see page 68) is something other than REAL.

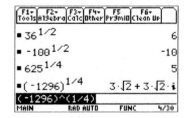

In fact, if the complex format setting is RECTANGULAR or POLAR, the TI-89 is a bit too clever to be useful! Observe the results reported for (e) and (f) in the screen shown here. These *are* correct, in a way; for example, the cube of $\frac{3}{2} + \frac{3\sqrt{3}}{2}i$ really is -27, but they are not the results that we are looking for.

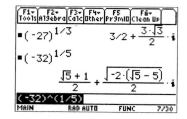

If we set the complex format to REAL, we get the desired results.

Section R.6   Example 5   (page 56)
Using the Definition of $a^{m/n}$

For (f), attempting to evaluate $(-4)^{5/2}$ will produce either a complex result or the error message shown, depending on the complex format setting (see the previous example).

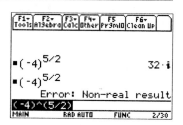

Section R.7   Example 1   (page 63)
Evaluating Roots

Aside from the square root function ∫ (2nd×), the TI-89 has no other root functions, so expressions like those in this example must be entered in exponential form.

Section R.7   Example 5   (page 65)
Using the Rules for Radicals to Simplify Radical Expressions

If the TI-89 is in AUTOMATIC or EXACT mode (see page 68), it can simplify many such expressions. Here are the TI-89's results for (a), (c), and (d).

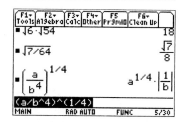

Section 1.1   Example 1   (page 85)
Solving a Linear Equation

The TI-89 offers several approaches to solving (or confirming a solution to) an equation. Aside from graphical approaches to solving equations (which are discussed later, on page 85), here are some non-graphical (numerical) approaches: As illustrated on the right, the TI-89's solve and nSolve functions attempt to find solutions to an equation, while the zeros function attempts to find the zeros of an expression. Full details on how to use these functions (all of which are found in the Algebra menu) can be found in the TI-89 manual.

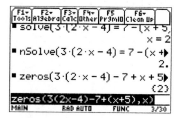

Additionally, the TI-89 includes an "interactive solver," accessed with APPS 9. This prompts for the equation to be solved, then allows the user to enter a guess for the solution (or a range or numbers between which a solution should be sought). To solve the equation, place the cursor on the line beginning with x= and press F2 (Solve).

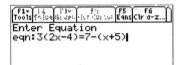

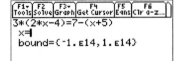

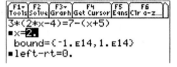

Enter equation   Specify guess, or press F2   Here is the solution.

The solver can also be used with equations containing more than one variable; simply provide values for all but one variable, then place the cursor on the line containing the variable for which a value is needed and press F2.

Note: In this example, we learned three methods by which the TI-89 can be used to support an analytic solution. But the TI-89 and any other graphing calculator also can be used for solving problems when an analytic solution is **not** possible—that is, when one cannot solve an equation "algebraically." This is often the case in many "real-life" applications, and is one of the best arguments for the use of graphing calculators.

Section 1.3   Example 1   (page 104)   Writing $\sqrt{-a}$ as $i\sqrt{a}$

Section 1.3   Example 2   (page 105)   Finding Products and Quotients Involving Negative Radicands

Section 1.3   Example 3   (page 105)   Simplifying a Quotient Involving a Negative Radicand

---

Note that the TI-89's Complex Format mode should be RECTANGULAR (see page 68). The screen on the right shows results from performing some computations in REAL mode; note that this *will* work for (a), (b), and (c) in Example 2, for which the final result is a real number, but an error occurs if the result is complex.

The screen on the right (also produced in REAL mode) illustrates an exception to this rule: A complex result produces no error in REAL mode if the entered expression included the character "ι" (2nd CATALOG —this is different from the "regular" lowercase i, alpha 9).

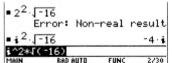

Use 2nd CATALOG for the "ι" character. Note that the TI-89's Complex Format mode should be REAL or RECTANGULAR (see page 68) to produce output similar to that shown in the text. The command ▶Frac ("to fraction"), shown in the screen accompanying Example 7, is not necessary (nor is it available) on a TI-89; provided it is set to either EXACT or AUTO mode, results will be displayed as exact fractions.

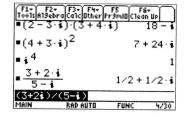

The solve and zeros functions, previously described on page 77, can be used for many types of equations, including quadratics and cubics like these. For equations with complex solutions, use cSolve or cZeros rather than solve and zeros.

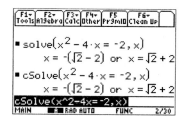
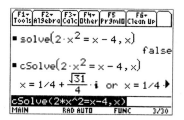
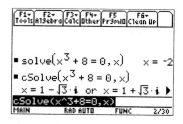

Section 1.5   Example 1   (page 122)   Solving a Problem Involving the Volume of a Box

A *table* can be a useful tool to solve equations like this. To use the table features of the TI-89, begin by entering the formula ($y = 15x^2 - 200x + 500$) on the Y= screen, as one would to create a graph. (The check marks determine which formulas will be displayed in the table, just as they do for graphs.)

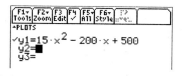

Next, press ◆ F4 to access the Table Setup screen. The table will display y values for given values of x. The tblStart value sets the lowest value of x, while Δtbl determines the "step size" for successive values of x. These two values are only used if the Independent option is set to AUTO—this means, "automatically generate the values of the independent variable (x)." The effect of setting this option to ASK is illustrated at the end of this example.

When the Table Setup options are set satisfactorily, press ENTER then ◆F5 to produce the table. The first screen on the right is the table generated based on the settings in the above screen; the fact that y is negative for $x = 8$ and $x = 9$ supports the restriction $x > 10$ from Step 3 of the example. By pressing ⊙ repeatedly, the x values are increased, and the y values updated. (Pressing ⊘ decreases the x values, but clearly that is not appropriate for this problem.) After pressing ⊙ several times, the table looks like the second screen, which shows that y1 equals 1435 when x equals 17.

Extension: Suppose the specifications called for the box to be 1500 cubic inches (instead of 1435). We see from the table above that x must be between 17 and 18 inches. The TI-89's table capabilities provide a convenient way to "zoom in" on the value of x (which could also be found with the solve feature, or by other methods).

Since $17 < x < 18$, set tblStart to 17 and Δtbl to 0.1. This produces the table shown on the right, from which we see that $17.2 < x < 17.3$. Now set tblStart to 17.2 and Δtbl to 0.01, which reveals that x is between 17.2 and 17.21. This process can be continued to achieve any desired degree of accuracy.

Finally, the screen on the right shows the effect of setting Independent to ASK. You enter an x value into the first column, and the corresponding y values will be computed.

Section 1.7 Example 1 (page 146)	Solving a Linear Inequality
Section 1.7 Example 2 (page 147)	Solving a Linear Inequality

The solve feature (described on page 77) will solve inequalities as well as equations. (2nd . types >; ◆0 types ≤; these symbols are also found in the MATH:Test menu.) While this is much faster, students should recognize that there are some benefits to doing problems like this by hand (the "hard" way)—specifically, working through the steps shown in the text gives one insight into what is going on, while using solve lends little to one's understanding of inequalities.

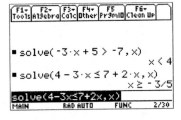

See also the discussion of Examples 5 through 9 (below) for another way to visualize the solutions to inequalities.

Section 1.7   Example 3   (page 148)   Solving a Three-Part Inequality

The `solve` feature will only handle one inequality at a time, so we must solve the two pieces separately (as shown) and then (optionally) join them with "and" (found in the MATH:Test menu). The final output shown is equivalent to $-\frac{7}{3} < x < 5$.

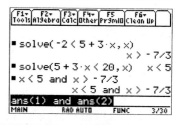

Section 1.7   Example 5   (page 149)   Solving a Quadratic Inequality
Section 1.7   Example 6   (page 150)   Solving a Quadratic Inequality
Section 1.7   Example 8   (page 152)   Solving a Rational Inequality
Section 1.7   Example 9   (page 153)   Solving a Rational Inequality

As the screen on the right shows, `solve` fails to give useful results for inequalities like these.

However, the TI-89's when function can be useful in visualizing solutions (as a way to check one's algebra). In the screen on the right, y1 was entered as `when(x^2-x-12<0,1,0)`, which means that y1 equals 1 whenever x satisfies this inequality.

In this window, we see that y1 is equals to 1 between -3 and 4, supporting the answer found in Example 5. Note that this picture does *not* help one determine what happens when $x = -3$ or $x = 4$; those two points must be checked separately.

Extension: Can the calculator be used to solve the inequality $x^2 + 6x + 9 > 0$? Yes—if it is used carefully. Shown is the graph of the expression y1=when(x^2+6x+9>0,1,0). This suggests that the inequality is true for all x; that is, the solution set is $(-\infty, \infty)$.

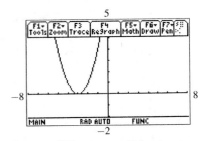

This conclusion is **not correct**: The graph of y2=x^2+6x+9 touches the *x*-axis at $x = -3$, which means that this must be excluded from the solution set, since "> 0" means the graph must be above (not on) the *x*-axis. The correct solution set is $(-\infty, -3) \cup (-3, \infty)$.

This example should serve as a warning: Sometimes, the TI-89 will mislead you. When possible, try to confirm the calculator's answers in some way.

Section 1.8 Example 1 (page 159)	Solving Absolute Value Equations
Section 1.8 Example 2 (page 160)	Solving Absolute Value Inequalities

The solve function described on page 77 of this manual can be used for these equations, too. Recall that abs(is found in the MATH:Number menu (2nd 5 1).

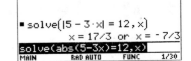

Section 2.1   Example 5   (page 186)   Using the Midpoint Formula

The TI-89 can do midpoint computations nicely by putting coordinates in a list—that is, using braces (2nd (and 2nd)) instead of parentheses. When adding two lists, the calculator simply adds corresponding elements, so the two *x*-coordinates are added, as are the *y*-coordinates. Dividing by 2 completes the task. (Note that, although the list is displayed in the history area as "{8   -4}", it was entered as "{8,-4}".)

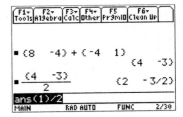

Section 2.2   Example 2   (page 194)   Graphing Circles

The text suggests graphing y1=4+√(36-(x+3)^2) and y2=4-√(36-(x+3)^2). Here are three options to speed up entering these formulas:

- After typing the formula in y1, move the cursor to y2, press ENTER, then press 2nd STO► (RCL), then type "y1" and press ENTER. This will "recall" the formula of y1, placing it on the entry line for y2. Now edit this formula, changing the first "+" to a "−."

- After typing the formula in y1, define y2=8-y1(x). This produces the desired results, since $8 - y1 = 8 - (4 + \sqrt{36 - (x+3)^2}) = 8 - 4 - \sqrt{36 - (x+3)^2} = 4 - \sqrt{36 - (x+3)^2}$.

- Enter the single formula y1=4+{-1,1}√(36-(x+3)^2). (The curly braces { and } are 2nd (and 2nd), respectively). When a list (like {-1,1}) appears in a formula, it tells the TI-89 to graph this formula several times, using each value in the list.

The window chosen in the text is a square window (see Section 11 of the introduction of this chapter), so that the graph looks like a circle. (On a non-square window, the graph would look like an ellipse—that is, a distorted circle.) Note, however, that on the TI-89, this window is not square. Other square windows would also produce a "true" circle, but some will leave gaps similar to those circles shown in Sections 11 and 12 of the introduction (pages 70–71). Shown is the same circle on the square window produced by "squaring up" the standard window.

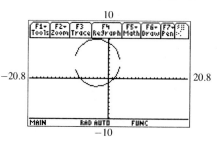

Section 2.3   Example 6   (page 208)   Using Function Notation

Section 2.3   Example 7   (page 209)   Using Function Notation

---

The TI-89 is quite good with function notation. In the screen shown, y1 has been defined as the function f from Example 6, and y2 as g from Example 7. It is then easy to compute $f(2)$ by entering y1(2), and (provided the variables a and q are undefined), we get the desired results from y1(q) and y2(a+1). The last entry shows that if a has a numerical value, the reported result is a number (rather than an expression). We can assure that all one-letter variables are undefined by pressing [2nd][F1][1].

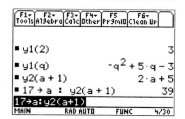

Section 2.4   Example 3   (page 219)   Graphing a Vertical Line

---

The text notes that such a line is not a function; therefore, it cannot be graphed by entering the equation on the Y= screen. However, the TI-89 does provide a fairly straightforward method of graphing vertical lines using the LineVert command.

From the home screen, type the LineVert command one letter at a time (upper/lower case is ignored), or using the function catalog (see page 66), then type the x-coordinate of the line. For example, LineVert -3 [ENTER] would draw the line $x = -3$ on the current window.

Alternatively, the vertical line can be drawn "interactively" using the GRAPH:Pen:Vertical command. This allows the user to move the cursor (and the vertical line) to the desired position using the ⊙ and ⊙ keys. Pressing [ENTER] places the line at that location, then allows more lines to be drawn. (Press [ESC] to finish drawing vertical lines.) Note that since, on the standard window, $x = -3$ does not correspond to a column of pixels on the screen (see page 70), the vertical line is actually located at $x \approx -3.038$, but the appearance on the screen is the same as if it were at $x = -3$. The process is illustrated below.

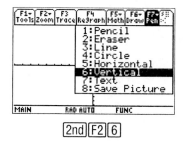

[2nd][F2][6]

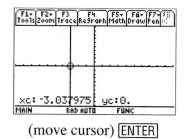

(move cursor) [ENTER]

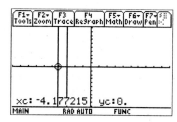

Make more lines, or [ESC]

Section 2.5   Example 6   (page 236)   Finding Equations of Parallel and Perpendicular Lines

In the discussion following this example (on page 237), the text notes that the lines from part (b) do not appear to be perpendicular unless they are plotted on a square window. See section 11 of the introduction of this chapter (page 70) for more information about square windows.

Section 2.5   Example 7   (page 238)   Finding an Equation of a Line That Models Data

Section 2.5   Example 8   (page 239)   Finding a Linear Equation That Models Data

Aside from the methods described in these two examples, the text refers to the technique of linear regression on page 241. Figure 51 illustrates the steps on a TI-83; here is a more detailed description of the process on a TI-89 (using the data of Example 8).

Given a set of data pairs (x, y), the TI-89 can produce a scatter diagram and can find various formulas (including linear and quadratic, as well as more complex formulas) that approximate the relationship between x and y. These formulas are called "regression formulas."

The first step in determining the regression formula is to enter the data into the TI-89. This is done by pressing [APPS][6][3] (Data/Matrix Editor), then entering a name for the "data variable" (which will contain all of the numbers for the regression). Alternatively, use an existing data variable, if there is one.

In the spreadsheet-like screen that appears, enter the year values into the first column (c1) and the percentage of women into the second column (c2). If re-using an existing data variable, old data can be cleared out using Tools:Clear Editor, Utils:Delete, or Utils:Clear Column.

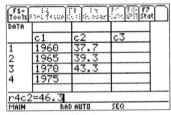

To find the line of best fit, press [F5] (Calc). For Calculation Type, choose LinReg, and set x and y as c1 and c2, respectively.

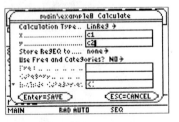

Pressing [ENTER] displays the results of the LinReg (shown on the right). These numbers agree (after rounding) with those in the text.

Section 2.5   Example 9   (page 241)   Solving an Equation with a Graphing Calculator

The text illustrates a graphical process for solving the equation $-2x - 4(2-x) = 3x + 4$; in this discussion, we will show two graphical methods for solving the equation $\frac{1}{2}x - 6 = \frac{3}{4}x - 9$. (The answer is $x = 12$.)

The first method is the **intersection** method. To begin, set up the TI-89 to graph the left side of the equation as y1, and the right side as y2. **Note:** Entering the fractions in parentheses—e.g., y1=(1/2)x-6—ensures no mistakes with order of operations. This is not crucial for the TI-89, but is a good practice because some other models give priority to implied multiplication. See section 8 of the introduction, page 67.

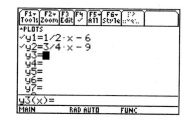

We are looking for an x value that will make the left and right sides of this equation equal to each other, which corresponds to the x-coordinate of the point of intersection of these two graphs.

Next, select a viewing window which shows the point of intersection; we use $[-15, 15] \times [-10, 10]$ for this example. The TI-89 can automatically locate this point using the GRAPH:Math:Intersection feature. Use $\odot$, $\ominus$ and ENTER to specify which two functions to use (in this case, the only two being displayed). The TI-89 then prompts for lower and upper bounds (numbers that are, respectively, less than and greater than the location of the intersection). After pressing ENTER, the TI-89 will try to find an intersection of the two graphs. The screens below illustrate these steps.

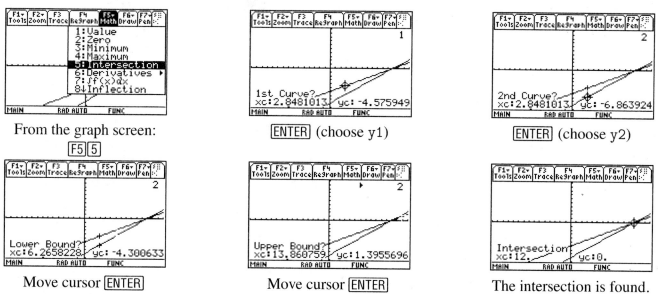

From the graph screen:
F5 5

ENTER (choose y1)

ENTER (choose y2)

Move cursor ENTER

Move cursor ENTER

The intersection is found.

The x-coordinate of this point of intersection is calculated to 14 digits of accuracy, so if the solution were some less "convenient" number (say, $\sqrt{3}$ or $1/\pi$), we would have an answer that would be accurate enough for nearly any computation.

Note: An approximation for the point of intersection can be found simply by moving the TRACE cursor as near the intersection as possible. The amount of error can be minimized by "zooming in" on the graph. This is the only method available for graphing calculators such as the TI-81.

The second graphical approach is to use the x-**intercept method**, which seeks the x-coordinate of the point where a graph crosses the x-axis. Specifically, we want to know where the graph of y1−y2 crosses the

x-axis, where y1 and y2 are as defined above. This is because the equation $\frac{1}{2}x - 6 = \frac{3}{4}x - 9$ can only be true when $\frac{1}{2}x - 6 - \left(\frac{3}{4}x - 9\right) = 0$. (This is the approach illustrated in the text for Example 9.)

To find this x-intercept, begin by defining y3=y1(x)−y2(x) on the Y= screen. We could do this by re-typing the formulas entered for y1 and y2, but having typed those formulas once, it is more efficient to do this as shown on the right. Note that y1 and y2 have been "de-selected" so that they will not be graphed (see section 12 of the introduction, page 71).

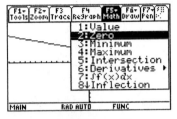

We must first select a viewing window which shows the x-intercept; we again use $[-15, 15] \times [-10, 10]$. The TI-89 can automatically locate this point with the GRAPH:Math:Zero feature. The TI-89 prompts for lower and upper bounds (numbers that are, respectively, less than and greater than the zero), then attempts to locate the zero between the given bounds. (Provided there is only one zero between the bounds, and the function is "well-behaved"— meaning it has some nice properties like continuity—the calculator will find it.) The screens below illustrate these steps.

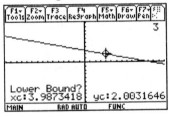

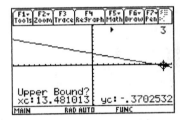

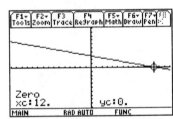

Move cursor to	Move cursor to	The TI-89
the left of the zero,	the right of the zero,	finds the zero.
press ENTER	press ENTER	

Section 2.6   Example 2   (page 252)   Graphing Piecewise-Defined Functions

The text shows how to enter piecewise-defined functions on the TI-83; for a TI-89, the appropriate way to enter the function in Example 2(a) is shown on the right. Note the difference between how the function is entered and how it is displayed; in particular, the display is similar to how the function appears in print. (Use ◆0 to type "≤".) The use of DOT mode is not crucial to the graphing process, as long as one remembers that vertical line segments connecting the "pieces" of the graph (in this case, at $x = 2$) are not really part of the graph.

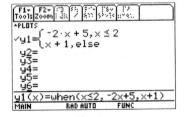

Extension: A more complicated piecewise-defined functions, such as

$$f(x) = \begin{cases} 4 - x^2 & \text{if } x < -1 \\ 2 + x & \text{if } -1 \le x \le 4 \\ -2 & \text{if } x > 4 \end{cases}$$

is entered as

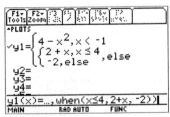

```
when(x<-1,4-x^2,when(x<=4,2+x,-2))
```

which the TI-89 displays as shown on the right. This tells the TI-89 to compute y1 as follows: (a) If $x < -1$, use the expression $4 - x^2$. (b) Otherwise (if $x \ge -1$), if $x \le 4$, use $2 + x$. (c) Otherwise, use -2.

Section 2.6   Example 3   (page 253)   Graphing a Greatest Integer Function

We wish to graph the step function $y = [[\frac{1}{2}x + 1]]$. On the TI-89, the greatest integer function is called floor or int; the former is found in the MATH:Number menu ([2nd][5][1][6]), while the latter is in the [CATALOG] (or either can be typed one letter at a time). Thus, either y1 or y2 (as shown on the right) will produce this graph.

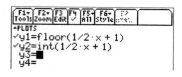

Section 2.7   Example 1   (page 259)   Stretching or Shrinking a Graph
Section 2.7   Example 2   (page 261)   Reflecting a Graph Across an Axis
Section 2.7   Example 6   (page 265)   Translating a Graph Vertically
Section 2.7   Example 7   (page 266)   Translating a Graph Horizontally

See page 72 for information on setting the thickness of a graph. Note that the symbol next to Y2 in the screen shown in the text indicates that the graph should be thick.

It is possible to distinguish between the two graphs without having them drawn using different styles. When two or more graphs are drawn, the TRACE feature can be used to determine which graphs correspond to which formulas: By pressing ⊙ and ⊙, the trace cursor jumps from one graph to another, and the upper right corner displays the number of the current formula. On the right, the trace cursor is on graph 2—that is, the graph of $g(x) = 2|x|$.

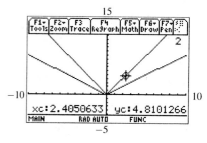

Section 2.8   Example 4   (page 278)   Finding the Difference Quotient

Because the TI-89 can do algebraic manipulations and simplifications with variables, it can find this difference quotient.

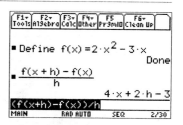

Section 2.8   Example 5   (page 279)   Evaluating Composite Functions
Section 2.8   Example 6   (page 280)   Determining Composite Functions and Their Domains
Section 2.8   Example 7   (page 280)   Determining Composite Functions and Their Domains

The expressions shown in the calculator screens on page 279 will also work on the TI-89's home screen. The command ▸Frac ("to fraction") is not necessary (nor is it available) on a TI-89; provided it is set to either EXACT or AUTO mode, results will be displayed as exact fractions.

The screen on the right illustrates how the TI-89 can perform these compositions (provided the variable x has no numerical value assigned to it). This screen shows the functions f and g being defined on the home screen; we could also define y1 and y2 on the Y= screen.

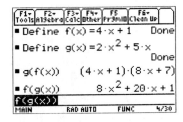

Section 3.1	Example 1	(page 304)	Graphing Quadratic Functions
Section 3.1	Example 2	(page 305)	Graphing a Parabola by Completing the Square
Section 3.1	Example 3	(page 306)	Graphing a Parabola by Completing the Square

---

The TI-89 can automatically locate extreme values ("hills" and "valleys") in a graph using the Minimum and Maximum options in the GRAPH:Math menu (F5 from the graph screen). We illustrate this procedure for the function in Example 1(c).

Enter the function in y1, and graph in a window that shows the extreme value. Since the coefficient of x^2 is negative, this is a parabola that opens down, and the extreme point is a maximum value. Press F5 4, then use the arrow keys and ENTER to define lower and upper bounds, as was done previously with the Math:Zero and Math:Intersection commands—see the discussion on page 85 of this manual.

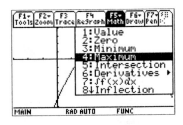

Depending on the window and the specified bounds, the x value may be off a bit from the exact answer, as is the case here. A limitation of the technology is that the calculating algorithms are programmed to stop within a certain degree of accuracy. It is important for the user to recognize this limitation and suspect slight errors when calculations are this close to integer values.

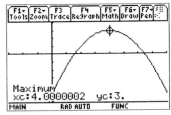

| Section 3.1 | Example 6 | (page 310) | Modeling the Number of Hospital Outpatient Visits |

See page 84 for information about using the TI-89 for regression computations.

| Section 3.2 | Example 3 | (page 325) | Deciding Whether a Number is a Zero |

The TI-89's "function notation" evaluation method (page 83 of this manual) can also be used to check for zeros: In the screen on the right, y1, y2, and y3 have been defined (respectively) as the functions in (a), (b), and (c).

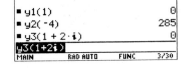

Section 3.3   Example 6   (page 335) — Finding All Zeros of a Polynomial Function Given One Zero
Section 3.4   Example 7   (page 349) — Approximating Real Zeros of a Polynomial Function

The TI-89's algebraic manipulation capabilities make problems like this very simple. The `factor(` and `cFactor(` commands are found in the Algebra menu (F2 2 on the home screen). Also found in that menu (but not shown here), the `zeros` and `cZeros` commands can produce a list of zeros (real only, or both real and complex) of the polynomial.

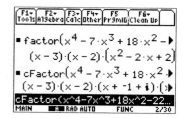

Additionally, if installed on your calculator (see section 13 of the introduction, page 73), the "Polynomial Root Finder" APP can find all zeros (real and complex) of a polynomial. The screens below illustrate the process.

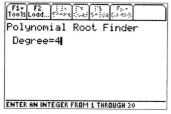

Give the degree of the polynomial.

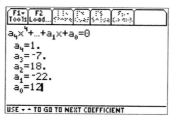

Enter coefficients on this screen.

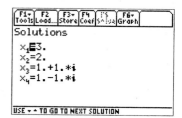

Press F5 to find the zeros.

Section 3.5   Example 2   (page 361) — Graphing a Rational Function

Note that this function is entered as y1=2/(1+x), **not** y1=2/1+x. (The TI-89's "pretty print" display makes this sort of mistake unlikely.)

The issue of incorrectly drawn asymptotes is also addressed in section 12 of the introduction (page 71). Changing the window to $xmin = -5$ and $xmax = 3$ (or any choice of $xmin$ and $xmax$ which has -1 halfway between them) eliminates this vertical line because it forces the TI-89 to attempt to evaluate the function at $x = -1$. Since f is not defined at -1, it cannot plot a point there, and as a result, it does not attempt to connect the dots across the "break" in the graph.

Section 4.1   Example 7   (page 409) — Finding the Inverse of a Function with a Restricted Domain

Note that Figure 10 shows a square viewing window (see page 70); the mirror-image property of the inverse function would not be as clear on a non-square window. Note also that although this window is square on a TI-83, it is not quite square on a TI-89.

Section 4.2   Example 11   (page 426) — Using Data to Model Exponential Growth

See page 84 for information about using the TI-89 for regression computations.

Section 4.4   Example 1   (page 448)   Finding pH

The natural logarithm function (ln) is [2nd][X]. For the common (base-10) logarithm, use the CATALOG or type log one letter at a time.

For (a), the text shows `-log(2.5*10^(-4))`. The TI-89 does not have a "`10^`" function, but [1][0][^] produces the same results. This could also be entered as shown on the first line of the screen on the right, since "`E`" (produced with [EE]) and "`*10^`" are nearly equivalent. The two are not completely interchangeable, however; in particular, in part (b), `10^(-7.1)` **cannot** be replaced with `E-7.1`, because "`E`" is only valid when followed by an *integer*. That is, `E-7` produces the same result as `10^-7`, but `E-7.1` is interpreted as `(1E-7)*0.1`.

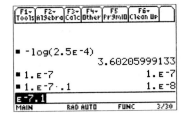

Section 4.5   Example 8   (page 463)   Modeling Coal Consumption in the U.S.

The modeling function given in the text was found using a regression procedure (LnReg, or logarithmic regression) similar to that described on page 84.

Section 5.1   Example 1   (page 495)   Solving a System by Substitution
Section 5.1   Example 3   (page 496)   Solving an Inconsistent System
Section 5.1   Example 4   (page 497)   Solving a System with Infinitely Many Solutions
Section 5.1   Example 6   (page 500)   Solving a System of Three Equations with Three Variables

If installed on your calculator (see section 13 of the introduction, page 73), the "Simultaneous Equation Solver" APP allows you to solve systems of linear equations. The screens below illustrate the process. When all coefficients have been entered (in the form of an *augmented matrix*, discussed in Section 5.2 of the text), pressing [F5] solves for the three unknowns (which the TI-89 calls x_1, x_2, and x_3, rather than x, y, and z).

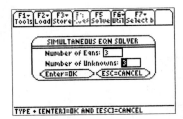

Give the number of
equations and unknowns.

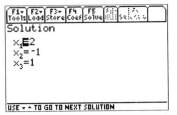

Enter coefficients on this
screen.

Press [F5] to solve the
system.

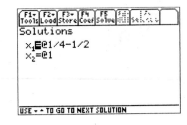

This APP reports "No solution found" for systems with no solution. For systems with infinitely many solutions, the APP gives results like those shown on the right (for Example 4). "@1" means "an arbitrary value," so this output means "$x = y/4 - 1/2$, and y can be anything"—the same as in the comment following Example 4. (Recall that the TI-89 uses variables x_1 and x_2, rather than x and y.)

Section 5.1   Example 8   (page 502)   Using Curve Fitting to Find an Equation Through Three Points

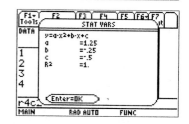

Note that the equation found in this example by algebraic methods can also be found very quickly on the TI-89 by performing a quadratic regression (QuadReg—see page 84) on the three given points. These agree with the coefficients found in the text.

Section 5.2   Example 1   (page 512)   Using the Gauss-Jordan Method
Section 5.2   Example 2   (page 514)   Using the Gauss-Jordan Method

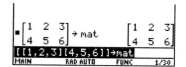

Before describing these operations, a few words about entering matrices into the TI-89. The matrices shown in the calculator screens in the text have the names "[A]" and "[B]." On the TI-82 and TI-83, all matrices have names of the form [*letter*]; on the TI-89, matrices can have any variable name. (In fact, variable names *cannot* include the bracket characters on the TI-89.) One way to set the values in a matrix is by "storing" the contents on the home screen; an example is shown on the right. (Note the difference between how this is *entered* and how it is *displayed* in the history area.) The square brackets are 2nd , and 2nd ÷ .

Alternatively, press APPS 6 (Data/Matrix editor) and either create a new matrix, or open an existing one. If creating a new matrix, you will be prompted for a name and the size of the matrix, as the screen on the right illustrates.

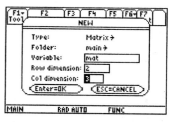

The matrix editor resembles a spreadsheet. Individual values of the matrix can be entered, or the matrix can be resized or sorted using the Util options (2nd F1). Press 2nd ESC (QUIT) when finished.

Here are the formats for the matrix row-operation commands (2nd 5 4 alpha ×), along with examples of their effect on the matrix $\text{mat}=\begin{bmatrix} 1 & 4 & 7 \\ 2 & 5 & 8 \\ 3 & 6 & 9 \end{bmatrix}$:

- rowSwap(*matrix*,*A*,*B*) produces a new matrix that has row *A* and row *B* swapped.

  Input: rowSwap(mat,1,2)   Output: $\begin{bmatrix} 2 & 5 & 8 \\ 1 & 4 & 7 \\ 3 & 6 & 9 \end{bmatrix}$

- rowAdd(*matrix*,*A*,*B*) produces a new matrix with row *A* added to row *B*.

  Input: rowAdd(mat,1,2)   Output: $\begin{bmatrix} 1 & 4 & 7 \\ 3 & 9 & 15 \\ 3 & 6 & 9 \end{bmatrix}$

- mRow(*number*, *matrix*, *A*) produces a new matrix with row *A* multiplied by *number*.

  Input: mRow(-4,mat,1)

  Output: $\begin{bmatrix} -4 & -16 & -28 \\ 2 & 5 & 8 \\ 3 & 6 & 9 \end{bmatrix}$

- mRowAdd(*number*, *matrix*, *A*, *B*) produces a new matrix with row *A* multiplied by *number* and added to row *B* (row *A* is unchanged).

  Input: mRowAdd(2,mat,1,3)

  Output: $\begin{bmatrix} 1 & 4 & 7 \\ 2 & 5 & 8 \\ 5 & 14 & 23 \end{bmatrix}$

Keep in mind that these row operations leave the matrix mat untouched. To perform a sequence of row operations, each result must either be stored in a matrix, or use the result variable ans(1) as the matrix. For example, with mat equal to the augmented matrix for the system given in this example, the command shown on the right takes that augmented matrix and multiplies the first row by 1/2 (the appropriate first step in the Gauss-Jordan method),

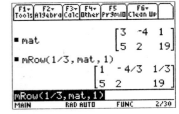

displaying the resulting matrix on the screen—but mat is exactly as it was before. Note the use of ans(1) in the three screens below, which complete the steps of the Gauss-Jordan method.

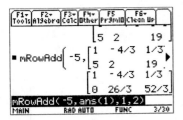 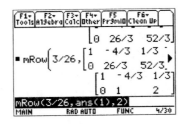 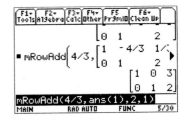

As a shortcut, the TI-89's rref(command (2nd 5 4 4) will do all the necessary row operations at once, making these individual steps seem tedious. However, doing the whole process step-by-step can be helpful in understanding how it works.

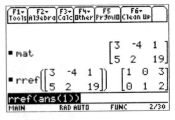

Section 5.3   Example 1   (page 523)   Evaluating a 2×2 Determinant

Section 5.3   Example 3   (page 526)   Evaluating a 3×3 Determinant

---

The determinant of any square matrix is easy to find with a TI-89. (At least, it is easy for the user; the calculator is doing all the work!) Simply choose the command det(from the MATH:Matrix menu (2nd 5 4 2), then type the matrix or the name of matrix variable. Note that trying to find the determinant of a non-square matrix (for example, a 3 × 4 matrix) results in a Dimension error.

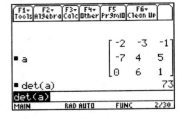

Section 5.4   Example 1   (page 536)

Finding a Partial Fraction Decomposition

The TI-89's expand(command (found in the Algebra menu, [F2][3]) automates the task of expanding a rational expression into partial fractions.

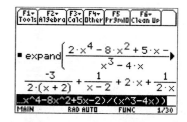

Section 5.5   Example 2   (page 544)

Solving a Nonlinear System by Elimination

See the discussion of Example 2 from Section 2.2 (page 82 of this manual) for tips on entering formulas like these.

The TI-89's solve function can handle system like this: Enter

    solve(x^2+y^2=4 and 2x^2-y^2=8,{x,y})

and the result is

    x=2 and y=0 or x=-2 and y=0

Section 5.6   Example 1   (page 555)

Graphing a Linear Inequality

With the TI-89, there are two ways to shade above or below a function. The simpler way is to use the Above and Below graph styles (see page 72), which mean "draw the graph of this function and shade the region above/below it." This style produces the graph shown below on the right. Note that the TI-89 is not capable of showing the detail that the line is "dashed."

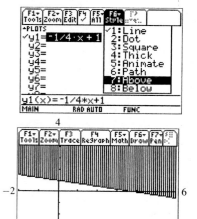

The other way to shade is the Shade command, accessed in the [CATALOG]. The format is

    Shade *lower*,*upper*,*min x*,*max x*,*pattern*,*resolution*

Here *lower* and *upper* are the functions between which the TI-89 will draw the shading (above *lower* and below *upper*). The last four options can be omitted. *min x* and *max x* specify the starting and ending *x* values for the shading. If omitted, the TI-89 uses xmin and xmax.

The last two options specify how the shading should look. *pattern* determines the direction of the shading: 1 (vertical—the default), 2 (horizontal), 3 (negative-slope 45°—that is, upper left to lower right), or 4 (positive-slope 45°—that is, lower left to upper right). *resolution* is an integer from 1 to 10 which specifies how dense the shading should be (1 = shade every column of pixels, 2 = shade every other column,

3 = shade every third column, etc.). If omitted, the TI-89 shades every other column; i.e., it uses *resolution* = 2.

To produce the graph shown above, the appropriate command (typed on the home screen) would be something like

```
Shade -(1/4)x+1,10.
```

The use of 10 for the *upper* function simply tells the TI-89 to shade up as high as necessary; this could be replaced by any number greater than 4 (the value of ymax for the viewing window shown). Also, if the function y1 had previously been defined as -(1/4)x+1, this command could be typed as Shade y1(x),10.

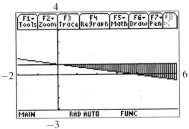

One more useful piece of information: Suppose one makes a mistake in typing the Shade command (e.g., switching *upper* and *lower*, or using the wrong value of *resolution*), resulting in the wrong shading. The screen on the right, for example, arose from typing Shade -(1/4)x+1,1. In order to achieve the desired results, the mistake must first be erased by pressing ⌨F4 (Regraph) or ⌨2nd⌨F1⌨1(Draw:ClrDraw). Then return to the home screen, correct the mistake in the Shade command, and try again.

Section 5.6   Example 2   (page 556)   Graphing Systems of Inequalities

The easiest way to produce (essentially) the same graph as that shown in the text is to use the "shade above/below" graph style (see page 72). The screen on the right (above) shows the style for y2 being set to Above; the style for y1 is also Above. The results are shown on the graph below. When more than one function is graphed with shading, the TI-89 rotates through the four shading patterns (see the previous example); that is, it graphs the first with vertical shading, the second with horizontal, and so on. All shading is done with a resolution of 2 (every other pixel).

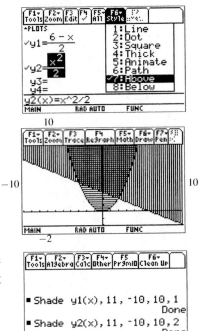

The Shade command (see the previous example) can be used to produce this from the home screen. If y1=(6-x)/2 and y2=x^2/2, the commands at right produce the graph shown above.

A nicer picture can be created, with a little more work, by making the observation that if y is greater than both $x^2/2$ and $(6-x)/2$, then for any x, y must be greater than the larger of these two expressions. The TI-89 provides a convenient way to find the larger of two numbers with the max(function, located in the MATH:List menu (2nd 5 3). Then max(y1(x),y2(x)) will return the larger of y1 and y2, and the Y= screen entries shown on the right (above) will produce the graph shown below it. (Note that y1 and y2 have been set to graph as solid curves, while y3 has the Above style.) The home-screen command Shade max(y1(x),y2(x)),11 would produce similar results.

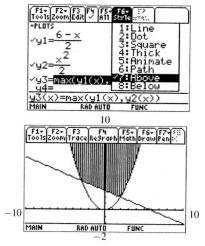

Extension: The table below shows how (using graph styles or home-screen commands) to shade regions that arise from variations on the inequalities in this example, assuming that y1=(6-x)/2 and y2=x^2/2. (The results of these commands are not shown here.)

For the system...	or equivalently...	the command would be...
$x < 6 - 2y$ $x^2 < 2y$	$y < (6 - x)/2$ $y > x^2/2$	shade below y1 and above y2, or enter Shade y2(x),y1(x)
$x > 6 - 2y$ $x^2 > 2y$	$y > (6 - x)/2$ $y < x^2/2$	shade above y1 and below y2, or enter Shade y1(x),y2(x)
$x < 6 - 2y$ $x^2 > 2y$	$y < (6 - x)/2$ $y < x^2/2$	shade below y1 and below y2, or enter Shade -10,min(y1(x),y2(x))

Section 5.7   Example 2   (page 566)   Adding Matrices

Section 5.7   Example 3   (page 567)   Subtracting Matrices

Section 5.7   Example 4   (page 568)   Multiplying Matrices by Scalars

Section 5.7   Example 6   (page 571)   Multiplying Matrices

See page 91 for information about matrices on the TI-89.

Section 5.8   Example 1   (page 580)   Verifying the Identity Property of I_3

On the TI-89, the function to produce identity matrices is identity(, located in the MATH:Matrix menu (2nd 5 4). The screen on the right shows the 3×3 identity matrix.

Section 5.8	Example 2	(page 583)	Finding the Inverse of a 3×3 Matrix
Section 5.8	Example 3	(page 584)	Identifying a Matrix with No Inverse

The TI-89 will quickly find the inverse of a square matrix (if it exists) by simply raising that matrix to the power −1 (that is, type the matrix, followed by ⌃ (−) 1). An example is shown. Attempting this with a non-square matrix results in a Dimension error. A Singular matrix error occurs when we try to find the inverse of this matrix. ("Singular" means noninvertible.)

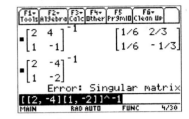

Section 6.1	Example 1	(page 607)	Graphing a Parabola with Horizontal Axis
Section 6.1	Example 2	(page 608)	Graphing a Parabola with Horizontal Axis
Section 6.2	Example 1	(page 618)	Graphing Ellipses Centered at the Origin
Section 6.3	Example 1	(page 629)	Using Asymptotes to Graph a Hyperbola

See the discussion of Example 2 from Section 2.2 (page 82 of this manual) for tips on entering formulas like these.

When entering these formulas, make sure to use enough parentheses, so that operations are performed in the correct order; e.g., for Example 2 of Section 6.1, the formula in y1 should look like this:

y1=-1.5+√((x-.5)/2)

(The TI-89's "pretty print" display should make it easy to notice a mistake in order of operations.)

Also, when looking at a calculator graph like the one shown in Figure 18 (page 619), keep in mind that the apparent gap between this graph and the *x*-axis is not really there; it is a flaw that arises from the calculator's method of plotting functions (see discussion on page 72 of this manual).

Section 7.1	Example 3	(page 655)	Modeling Insect Population Growth

One way to produce this sequence (and similar recursively defined sequences) is with the ans(1) variable, as shown on the right. After pressing 1 ENTER, then typing the next line and pressing ENTER, we simply press ENTER over and over to generate successive terms in the sequence. Each time, ans(1) is replaced by the previous result.

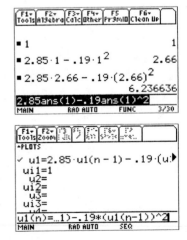

Figure 4 of the text shows a table of values. A similar table can be created by putting the TI-89 into Sequence mode, and making the definitions shown on the right. (This screen is the Y= screen, accessed with ♦ F1.) The plot in Figure 5(b) was also created in this way; see the discussion of Example 7 from Section 7.3 (below) for details.

| Section 7.1   Example 4   (page 657) | Using Summation Notation |
| Section 7.2   Example 9   (page 669) | Using Summation Notation |

The seq(command can be found in the MATH:List menu ([2nd][5][3]). Given a formula a_n for the nth term in a sequence, the command

$$\text{seq}(\textit{formula}, \textit{variable}, \textit{start}, \textit{end}, \textit{step})$$

produces the list $\{a_{start}, a_{start+step}, \ldots, a_{end}\}$. If *step* is omitted, a value of 1 is assumed. The size of the resulting list is limited by available memory.

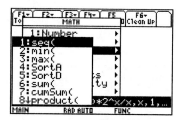

Note that *variable* can be any letter (or letters)—K and I are used in the text, but x, y, z, or t would be more convenient (since they can be typed with a single key).

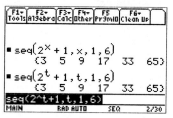

The sum(command is also found in the MATH:List menu ([2nd][5][3]). sum can be applied to any list—either to a list variable, or directly to a list created with the seq(command. Evaluating summations on the TI-89 requires first generating the sequence as a list, then summing the list. The summation in Example 4 can be performed as shown on the right, using the ans(1) storage variable.

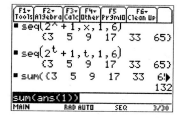

Alternatively, the sum(and seq(commands can be combined on a single line. The screen on the right uses this approach to find the answer for Example 9(a).

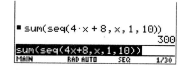

The Σ function, found in the Calc menu ([F3][4]), essentially combines the actions of the sum and seq commands. It has the added benefit that the entry in the history area looks the same as in the text.

| Section 7.3   Example 7   (page 676) | Summing the Terms of an Infinite Geometric Series |

The plot shown in the text can be produced in at least two ways. The first method involves putting the TI-89 in Sequence mode, then entering the formula for S_n on the u1= line of the Y= screen, as on the right. Also, it is not necessary to specify the value of ui1 (the initial term in the sequence); this is only needed for recursively-defined sequences.

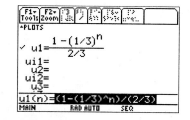

Next, press [♦][F2] and make the settings shown on the right. The viewing window is as shown in the text (xmin = 0, xmax = 6, ymin = 0, ymax = 2). Pressing [♦][F3] should produce a plot like that shown in the text. (If it does not, press [♦][F1][2nd][F2] and check that Axes is set to TIME.)

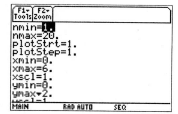

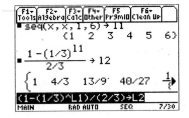

The second method begins by storing lists in two variables (we have used L1 and L2) as shown on the right. Note that this approach is less appealing for recursively defined sequences.

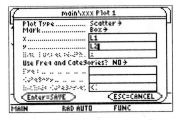

Next press ◆ F1 ⊙ ENTER and make the settings for a statistics plot as shown on the right. Finally, set up the viewing window, and check that nothing else will be plotted; that is, go to the Y= screen and make sure that the only Plot1 has a check mark next to it. Pressing ◆ F3 should produce a plot like that shown in the text.

Note: When finished with a statistics plot like this one, it is a good idea to turn it off so that the TI-89 will not attempt to display it the next time ◆ F3 (GRAPH) is pushed. This can be done from the Y= screen using F4 to un-check the plot, or by pressing F5 5 (All:Data Plots Off).

Section 7.4   Example 1   (page 687)   Evaluating Binomial Coefficients

Section 7.6   Example 4   (page 701)   Using the Permutations Formula

---

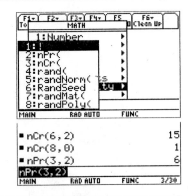

The nCr(and nPr(functions and the factorial operator "!" are found in the MATH:Probability menu (2nd 5 7). Note, though, that "!" is much more easily typed as ◆ ÷ —this is one of those "hidden" key combinations that is revealed by ◆ EE. The TI-89's format for nCr and nPr is different from the TI-83's (shown in the text); instead, enter these functions as shown on the right.

Section 7.7   Example 6   (page 716)   Finding Probabilities in a Binomial Experiment

---

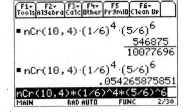

The TI-89 does not have statistical distribution functions like binompdf. The computations shown in this example must be done by manually entering the entire binomial probability formula (or by obtaining a program to automate such computations.) Note that if the TI-89 is in AUTO or EXACT mode, the results can be more detailed than is useful.